Algrove Publishing Limited
36 Mill Street, P.O. Box 1238
Almonte, Ontario, Canada K0A 1A0

Telephone: (613) 256-0350
Fax: (613) 256-0360
Email: sales@algrove.com
Web: www.algrove.com

Library and Archives Canada Cataloguing in Publication

Radford's practical barn plans : being a complete collection of practical, economical and common-sense plans of barns, out buildings and stock sheds / edited by William A. Radford.

(Classic reprint series (Almonte, Ont.))
Reprint. Originally published: Chicago : The Radord Architectural Co., 1909
Includes index.
ISBN 978-1-897030-65-3

1. Farm buildings--Designs and plans. 2. Barns--Designs and plans.
I. Radford, William A., 1865- II. Series.

NA8230.R324 2007 728'.922 C2007-906410-8

Printed in Canada
#1-11-07

Publisher's Note

The Radford Architectural Company was a major American supplier of building plans in the early 1900s. This is the second half of a book originally published as *Radford's Combined House and Barn Plan Book*. We left it at the original pagination (starting at page 115) so that the text references to various pages did not have to be revised, a process that is not only time consuming but can be a source of inadvertent errors.

Originally designed as a teaser to sell full sets of plans, it is still useful as a general guide to the overall design of basic farm structures.

Leonard G. Lee, Publisher
Almonte, Ontario
November 2007

RADFORD'S PRACTICAL BARN PLANS

Being a Complete Collection of Practical, Economical and Common-Sense Plans of Barns, Out Buildings and Stock Sheds.

EDITED BY

WILLIAM A. RADFORD

President of The Radford Architectural Co., Author of "Radford's Cyclopedia on Carpentry, Building and Architecture," "The Steel Square and Its Uses," "Practical Carpentry," "The Radford American Homes," "The Radford Ideal Homes," "Radford's Modern Homes," "Radford's Artistic Bungalows," "Radford's Artistic Homes," "Cement Houses and How to Build Them," "Radford's Combined House and Barn Plan Book," "Radford's Stores and Flat Buildings," and the best authority in the country on all matters pertaining to the building industry.

Originally published in 1909.

Algrove Publishing
Classic Reprint Series

PREFACE

The farm building department is given not only in the interest of the farmer, but because every one is more or less dependent upon the soil and is consequently directly or indirectly interested in the prosperity of the farmer. There comes a time when every business man and most other men want to build either a house, or barn, or both. It is the province of this book to offer suggestions. Plans are necessary to avoid mistakes. It is just as easy to secure good plans as to build after the ideas prevailing in the neighborhood, probably advanced by some carpenter who has had little or no experience outside of his own town. Barns as well as houses are built or should be built for a specific purpose and there are fundamental principles which are vital if the best results are to be obtained. It is not always necessary or desirable to build expensively. It often happens that an inexpensive or even a cheap structure will answer the purpose just as well, but this book does not countenance the building of unsightly houses or out buildings. There are essentials to be built into farm buildings that cannot be seen after they are finished, but these essentials must be there or the buildings will not be right. Drainage, foundation, ventilation and economy are four requisites that have received special attention in compiling the material for this work and they are four of the worst neglected principles when ordinary farm buildings are planned. Drainage and foundation are easily treated to fit each case, but ventilation is less tangible, although it is equally important, while the study of economy has no beginning and no legitimate ending for it embraces not only the construction of the building, but the use of it ever after. Even a small farm building is worth very close attention in the planning, location and building because of its appearance and the labor it entails in its connection with general farm economy.

In the arrangement of the book the first point made clear to the reader is the need for the structure. Naturally what the building is wanted for should be first made plain. Then the construction is followed to completion when the building takes its place in the farm economy.

It is a book, the author believes, which will make interesting a subject that farmers have always before them, and enable them to find a design for any kind of barn or farm building that may be needed. It is a pleasure to contribute anything to add to the beauties and charm of American farm life, and in that spirit this volume is given to the agricultural world.

Wm. A. Radford

Table of Contents

Please note that the page numbering in this book starts with page 115.

Department of Dairy Barns

Department of General Farm Barns

Department of Horse Barns

Department of Dairy Barns

LARGE DAIRY STABLE—A100

Price of Blue Prints, $15.00.

THE careful housing of dairy cows is receiving systematic consideration as never before. Investigations have been conducted by men who are thoroughly conversant with the subject from a practical as well as a scientific standpoint.

Government milk inspectors, backed by public opinion, have established a thorough system of inspection. City milk supply is now traced to its source, the cows examined thoroughly for condition and health and the stable for cleanliness. If incompetency or indifference has led the dairyman to disobey the state sanitary requirements he is not permitted to ship milk until he satisfies the inspector that he has mended his ways. This course was made necessary by the rapidly increasing volume of business which is conducted by such a cosmopolitan class of people; comprising as it does, all grades of producers from the most progressive farmer down the line of small dairymen to the ignorant huckster. Cleanliness is required by inspectors, first, last and all the time; thus, making the right start, for cleanliness, leads to many virtues. A man who is particular about all utensils, his wagon, stable, cattle and himslf, will not tolerate a poor stable or an unhealthy cow. He may not understand the science of ferments or disease germs, but his milk supply will be good and wholesome, because he robs harmful bacteria of the dirt upon which they thrive.

In our northern climate, warmer stables have for years occupied the attention of our best farmers and stockmen. Bank barns were the outgrowth of a desire to provide comfortable stables that were both warmer and better. The convenience of having all stock under one roof, tucked carefully away from the cold, with plenty of feed overhead ready at all times to find its way to mangers and food racks by gravity, proved very alluring to ambitious farmers all over the country. But animals housed in these expensive dungeons were not happy and showed their discomfiture in watery eyes, lusterless hair, hot noses and hot, feverish breath, with fretful, quarrelsome actions together with their inability to grow or fatten. Too frequently cattle thus housed were attacked by bovine disease germs, which were materially assisted in their work of destruction by conditions so expensively though unintentionally provided. Stockmen thought the trouble was caused by too great a change in temperature by allowing the cattle to go out for an airing or for water each day; to remedy this, water buckets were added to the stable outfit and the stock confined in an abominable atmosphere for weeks at a time.

Atmospheric conditions affect animals differently. The heavy breeds of beef cattle are usually phlegmatic in disposition, paying little attention to ordinary disturbances; these suffered less in consequence, though it was noticed that they did not benefit from the quantity and quality of feed as they should. Milch cows of a highly nervous organization are more susceptible to incipient diseases caused by objectional surroundings than any other domestic animal. Not until progressive scientific men spent much time and money in investigations and experiments was the trouble traced to its true source.

Analyzing stable atmosphere led to the detection of harmful bacteria in incredulous numbers. Scientists engaged in the work were slow to give out the result of their first investigations, thinking that the conditions under which they were working might be abnormal. Prospecting further

and while endeavoring to learn the cause they found the conditions in these cellar stables particularly favorable to the propagation of stockmen's worst enemy. Harmful bacteria delight in a dusty atmosphere, especially when it is impregnated with moisture; when a share of the dampness comes from the moisture laden breath of animals that are obliged to breathe the same air over and over again, bacteria conditions are complete.

Bank barns are always damp and always dusty; owing to their construction they never admit sunlight in quantities sufficient to be of use. Sunlight is destructive to all forms of harmful bacteria; therefore a stable should admit the direct rays of the sun to every stall if possible.

An eastern model dairy stable combining all good qualities while eliminating objectionable features is shown in the accompanying plans. The stable may be built at a low cost, is warm in winter, cool in summer, and sanitary and hygienic at all times.

Location.

The proper location for a dairy stable is the first consideration. Good air, good drainage, plenty of sunlight and an abundant water supply are all essential features Fresh air and drainage may be secured by selecting an elevation; protection from cold winds by means of a tree belt or a high tight board fence. Sufficient water may be obtained in most any situation by a powerful windmill. There are other considerations such as convenience to the pasture fields and a short haul from the fields in which soiling crops are grown. Pasture, however, receives less consideration than it did a few years ago. North of parallel 42 there is an average of only six weeks of good pasture. Summer droughts sandwiched in between late spring and early fall frosts are responsible for this condition, so that a good many farmers in the east depend upon soiling crops a great deal more than they do on pasture. A runway consisting of least a quarter of an acre for each cow is necessary, but the fields may be more profitably employed in raising cultivated crops. The question of drainage is a very important one. If the soil is naturally dry and slopes sufficient to carry off rain water no elaborate system of tiling will be necessary, but if there is any doubt it is better to be on the safe side.

Grading.

In laying out a stable a great deal of after work may be saved by a careful survey of the grade. Manure should be removed from a dairy stable promptly every day and carted at once to the fields. By the use of a manure carrier and a spreader this way of managing is cheaper as well as better than the old fashioned way of piling in manure to be hauled away at some future time. In making the grade the stable floor may be placed high enough to run the manure carrier directly out over the spreader. Calculation must also be made for carrying off the water used in flushing the gutters and in washing the dairy utensils. The intake for ventilation is another consideration before commencing work. In order to lay out the ground right a general working drawing giving the floor plan and profile is necessary. Any one can work to such a plan by having a few simple instruments. An A level and a few stakes of different lengths comprises about all the tools necessary.

Excavation.

The excavation for the walls may be just deep enough to go below frost. For concrete or cement walls make the trench just the width necessary to hold the wall material but after the trench is done make a rounded recess all round the edge near the bottom to hold a course of three inch tile. This answers the double purpose of carrying off surplus water and preventing rats from undermining the wall. Rats will dig down at the side of the wall until they come to an obstruction, they will follow the obstruction along close to the wall but never think of digging outward to get around it. The ends of the tile should terminate in the main drain just below the trap.

Walls.

In some parts of the country stone is plentiful and farmers prefer to lay up a stone wall but generally speaking a concrete wall is cheaper and better. The materials may be put together on the ground and dumped into the trenches with unskilled labor. It is only necessary to look carefully to the leveling and finishing of the job. For this purpose a two inch plank staked carefully in position with the edges even with the top of the wall forms a guide both for leveling and for thickness. Openings in the plank may be left for doorways and boxes built around the size and shape to properly hold cement sills so that when the wall is finished the door sills will be

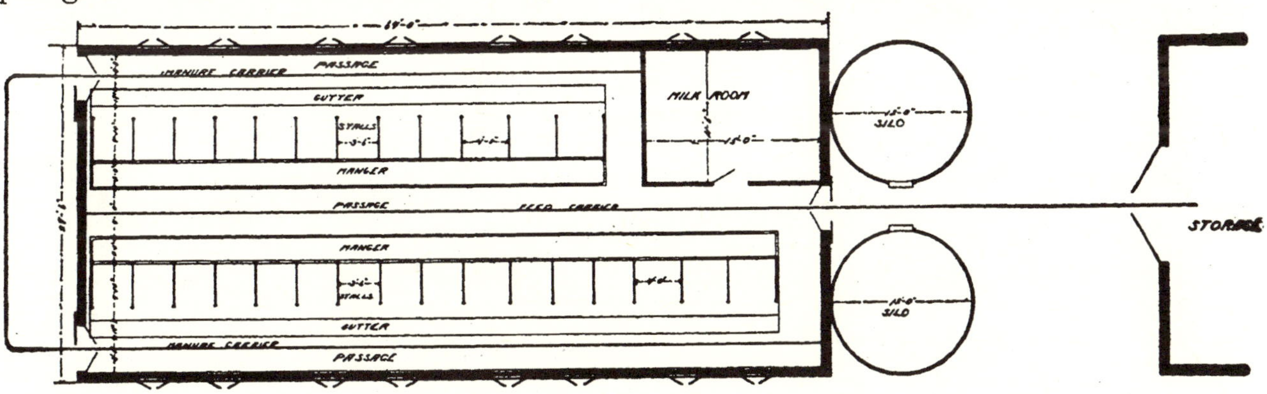

PLAN OF COW BARN

complete and the whole thing will be in one piece.

The Floor

After the walls are finished the grading for the floor comes next in order. The profile shows the relative position of the intake for fresh air, the floor of the feeding alley, position of the cement mangers, inclines of the floor in which the cattle stand, the gutter and the walk behind the cows. Besides the cross section the mangers and gutters incline with the length of the stable. In order to locate all these points a good many grade stakes are necessary. They are set carefully to measurement and driven down until the tops come right for the grade. It is easier to do this work before the building is erected. One point to be rembered is that the wall should not extend much above the floor for the reason that dampness will collect on the inner side or warmer side of the wall especially in winter. Also the iron pipes designed to partition the stalls and support the ceiling should be imbedded in cement when it is fresh.

Superstructure

It is cheaper to build barns and stables low because lighter material may be used in their construction. A dairy stable should have a low ceiling to facilitate ventilation. Seven feet is high enough for a ceiling but eight feet looks better if the stable is long and where there are a good many cows to-

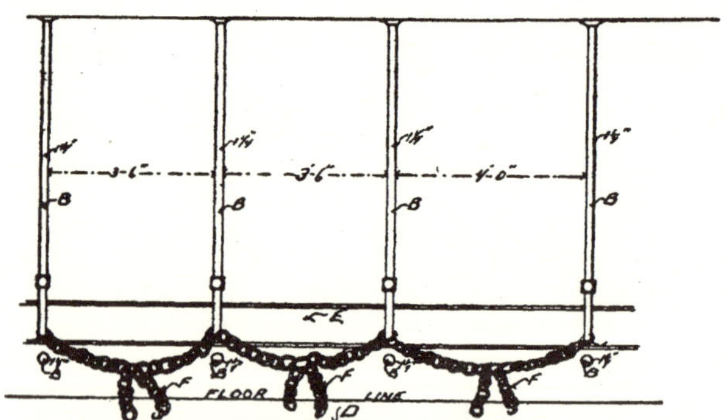

gether there is no objection to an eight foot ceiling. A good deal depends on the number of cows kept. A stable may be built on this plan to hold twenty-four cows or it may be made long enough to hold one hundred. The principle of ventilation depends on the circulation of air. Warm air is lighter than cold air and it naturally goes up. In order to ventilate a stable we must get animals enough in it to warm the air. There is little or no circulation in a cold room. For the ventilation to work right the temperature in a stable should

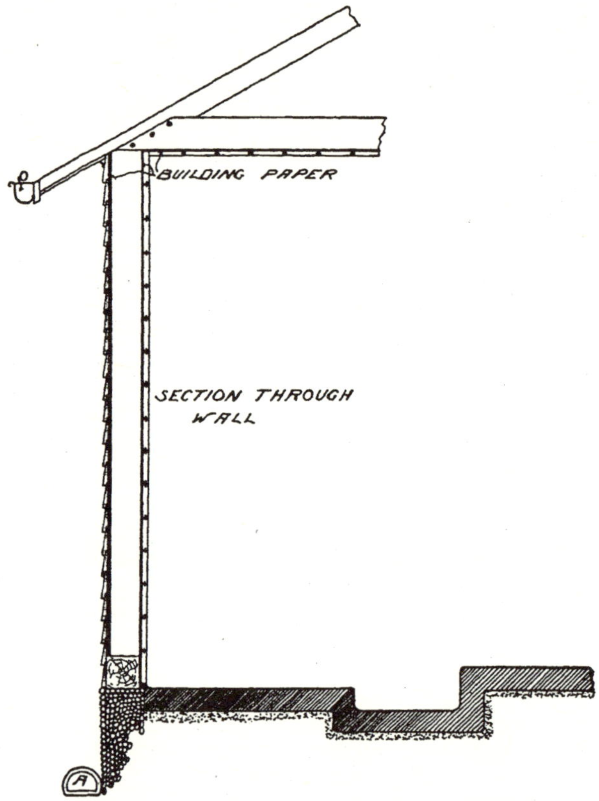

not go below 55 degrees. This plan takes the air in at the center in front of the cows where the cows may breathe the clean fresh air from outside before it becomes contaminated. The hot breath of the cows goes to the ceiling, spreads in all directions to the sides of the room while it loads up with impurities and finally settles to the floor at the sides of the stable where it is drawn off by the ventilators and sent out through the roof. In order for the ventilating system to work right the stable must be practically air tight around the sides and ceiling and the doors must fit well. There is a light sill six by six bedded in fresh cement mortar on top of the walls, two by six studding seven feet long toenailed into the sill and a two by six plate spiked on top of the studding. Building paper is nailed to the studding both inside and out. The inside is lined with matched ceiling without bead. This is to eliminate all cracks and joints as far as possible. There are no cracks and places for dust to lodge as all stable dust is bacteria laden. In like manner building paper is tacked to the ceiling joists and under the paper a light matched ceiling is nailed so that the whole room is smooth around and there are no projections or shelves of any kind to hold dust. The stall partitions are as light as possible for the same reason. Door and window frames are made flush on the inside and just a light four inch casing turned to cover the joint. It is better to use a great deal of care in laying the building paper around all such places to prevent air openings. It is not intended to use the loft over this stable for storage or any purpose but it is better to build the loft so that it may be swept occassionally to clear out the dust. A window is placed in each gable for the purpose of causing sufficient ventilation to keep the loft cool. The outside of the stable is boarded up with patent siding and a light box cornice makes the finish at the eaves. The ventilating system is shown in the cuts. It pays to to put on an eave trough whether the water is wanted for use or not because the drip from the eaves will cause dampness and this should be avoided. Because the building is low a light roof is sufficient. Two by four rafters are heavy enough if well supported by cross collar beams.

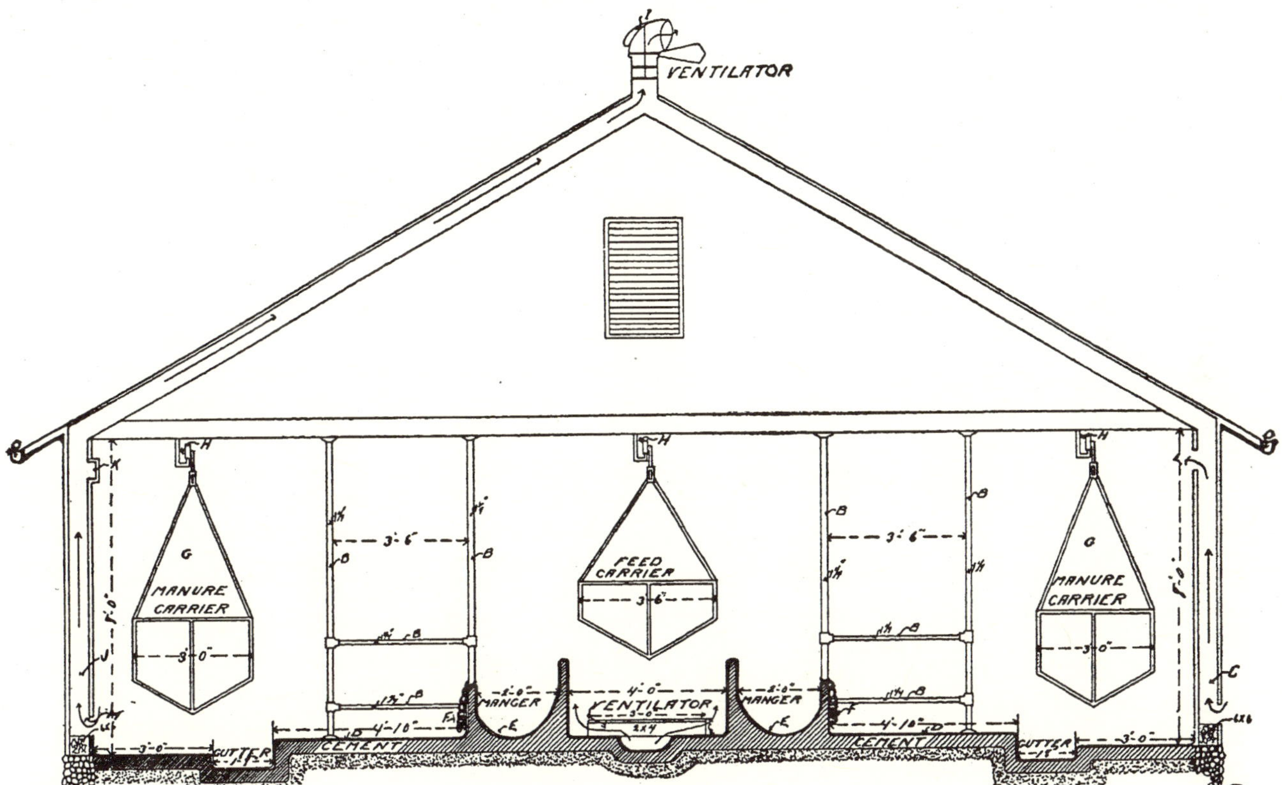

CROSS SECTION

The Silos

In this plan the silos are placed at the end of the stable. If the stable is long however it is better to put the silos in the middle. It will save steps at feeding time. It is better to have two small silos than one large one. From sixteen to twenty feet in diameter is big enough for any silo. The surface may then be fed off every day and the silage kept fresh at all times. The milk room is at the side of the silo. The floor and sides are built entirely of cement and the room has a white matched ceiling. It is provided with an open drain that connects with the main drain outside of the building. The milk room contains a separator, scales, Babcock tester and a shelf to hold the smaller utensils and a porcelain lined sink for washing dishes. Outside of the milk room is a rack to hold the cans where they are turned upside down every morning in the sun. Beyond the silos and milk room is the barn where the roughage is kept and the track from the stable runs across so the feed may be brought by an overhead track carrier. The silos are at the north end of the building. The manure is taken out through the south doors. The cows are also let in and out of the south doors. This style of stable should be built north and south so that the sun will shine in at all of the windows.

Silo Construction

The cheapest form of a silo is the round stave construction. It is about as good as any, too, when it is thoroughly weil built from well seasoned lumber; in fact, it has been thoroughly demonstrated that the stave silo is a success. In New Jersey and Eastern Pennsylvania the stave silo is almost universally used. They do not last as long as some others. Probably the average life of a stave silo is somewhere between five and ten years. But a farmer can tear down and rebuild because the material is comparatively cheap and there is not much of it. In some parts of the country there is a prejudice against this form of silo. Some claim that the silage is not so good, but it would be difficult to substantiate this claim. Of course, to keep silage properly in any kind of a silo it must be air tight. If a stave silo leaks at the joints the silage will suffer, but the same may be said of any make of silo.

Some of this prejudice comes from the dairy farmers who formerly had experience with stave silos which were constructed by putting rough planks together without beveling the edges, but the way staves are made now with bevels carefully cut to fit the circle and provided with heavy iron hoops, and plenty of them, there is probably no better construction. Some stave silos have round tongues and grooves. This is better than a plain straight bevel, but it is not absolutely necessary. The ends of the staves where they butt together are fitted with an iron tongue let into a saw cut in each end of the abutting staves.

A convenient height for a silo of this kind is thirty-two feet made from sixteen foot stuff, but some staves must be eight feet long in order to break joints. Most stave silos erected are bought from some manufacturer who has a patent on some little contrivance in connection with their manufacture, but any farmer can order the material and build his own silo if he wishes to do so. The mills will cut and bevel the staves and tongue and groove them to fit any circle desired, but it is necessary to understand all the little details and see that they are properly worked out. A good many of the patent silos have an iron framework to hold the doors. This is an advantage inasmuch as wood gets damp and swells, but any carpenter can bolt two timbers together in such a way as to make a good framework to hold the doors, and the saving in expense is considerable. The doors may be made loose and calked around the edges with tow or the soft parts of corn stalks makes very good calking material. In fact, there are a great many different ways to manage if a person is determined to have a silo, but it is well to remember that the doors are a particular part. The framework must be solid and there must be ample space between the doors for the hoops.

Figures on all Cuts Correspond.

A—Drain tile.

B—Gas pipe 1¼ inch for stall partitions, chain ring and ceiling supports.

C—Ventilation intake.

D—Stable floor where cows stand having an incline of two inches.

E—Cement manger having an incline of ⅛ inch per ten feet.

F—Cow chains.

G—Manure carriers.

H—Car tracks, should be near the ceiling to give plenty of head room.

I—Hood ventilator, tail on opening side with counter weight to prevent friction, and allow it to turn easily. This hood does not touch the pipe but turns on a spindle which passes through the upper cross piece in the pipe and is socketed in the lower cross piece about three feet down in the pipe.

J—Ventilator shaft drawing foul air from near the floor.

K—Register for use in hot weather to draw off the hot air when the stable doors are open.

L—Register that may be partially closed to regulate intake of fresh air.

M—Register to regulate the amount of draft allowed to foul air. This is one of the most important features of the system as the warmth of the stable as well as the quality of the air is controlled by it.

O—Galvanized iron gutter.

DAIRY BANK BARN—A125

An old fashioned dairy barn is shown in plan (A125). There are a good many such barns still in use in Wisconsin. Those using them say they are satisfactory under certain conditions.

LONGITUDINAL SECTION

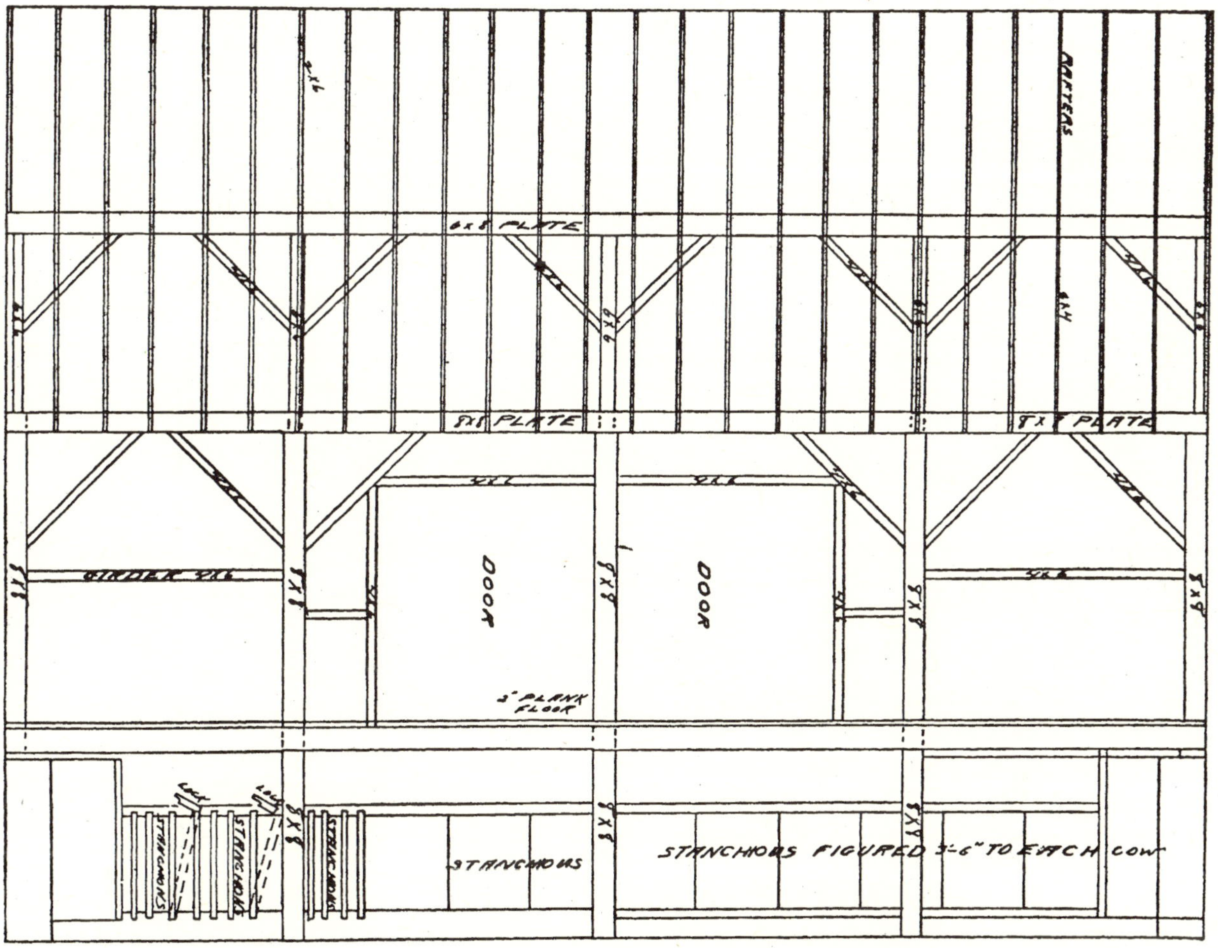

One good feature about this stable is the ventilation. To have good air in a cow stable it is absolutely necessary to have a system of ventilation. You can stable four or five cows together and depend on chance openings to provide them with oxygen, but you cannot depend on Providence to keep your cattle alive in a large stable unless you assist a little bit.

A good many dairymen prefer to have the cows face outward. This is a matter of individual preference. Probably nine stables out of ten are made to face the cows in, but this is no dead open and shut reason why this stable should be built that way. One advantage of having the two manure gutters in the middle is that a cart may be driven through to remove the manure. If there is any other good reason I am not familiar with it. In these Wisconsin stables the old fashioned stanchions are used.

There is a large amount of storage over-

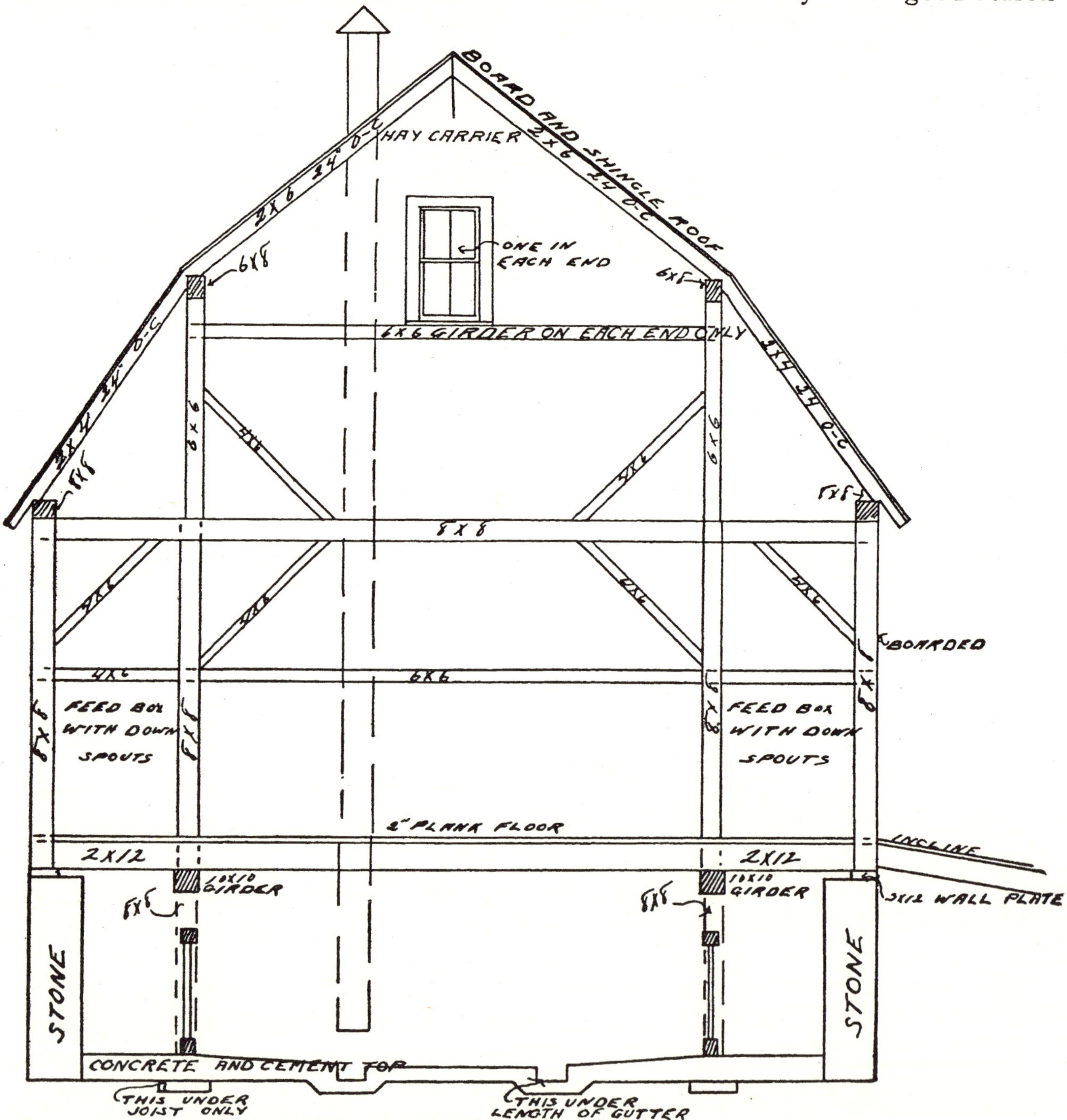

END SECTION

head in a barn like this, and it is a convenient barn to do the work in except in the matter of feeding the cows. It takes more steps to get around to feed the cows when they face out. This barn is backed up to a bank, preferably on the north side, where the incline may be had easily to drive in on the main floor. The horse fork is worked from the center.

Although stone is stipulated for the walls of this building, concrete of course will answer the purpose just as well, in fact concrete is better than stone when it is properly made. Concrete is damp-proof when the materials are so mixed that the sand fills the spaces between the broken rock and the cement fills the spaces between the grains of sand. Then if the mixing is thoroughly done and just enough water used to thoroughly amalgamate the different ingredients into one solid mass it will make a wall that dampness cannot penetrate.

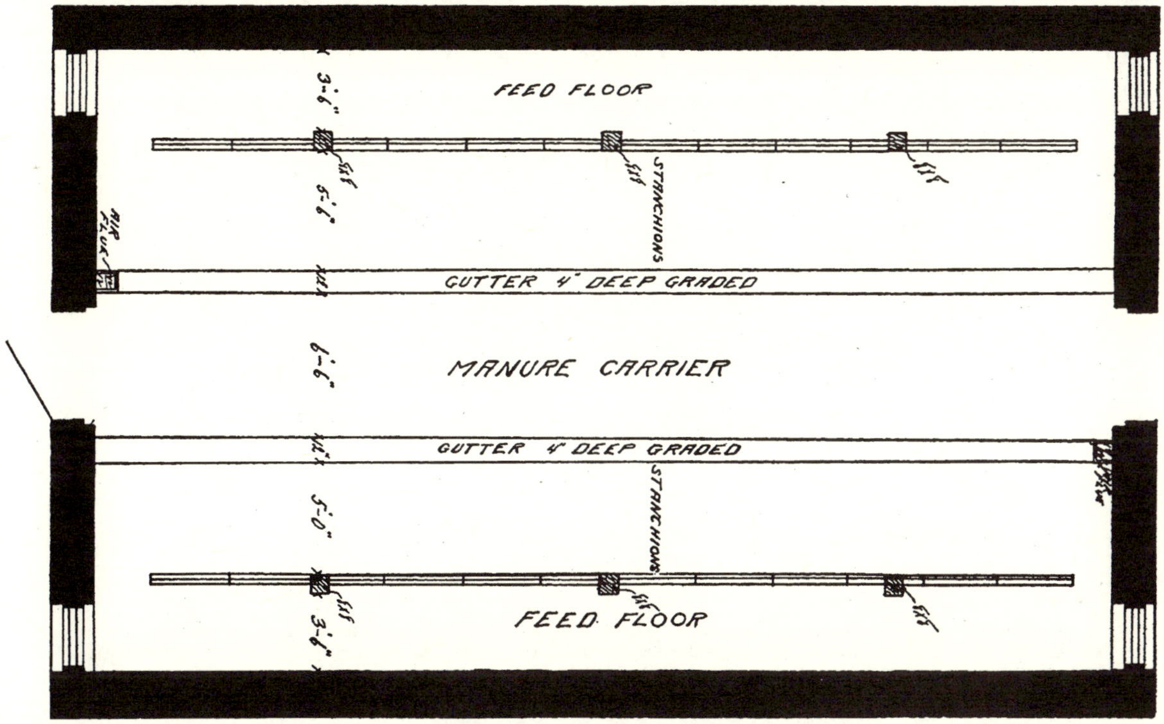

BARN AND COW STABLE 32 FT. WIDE 50 FT. LONG

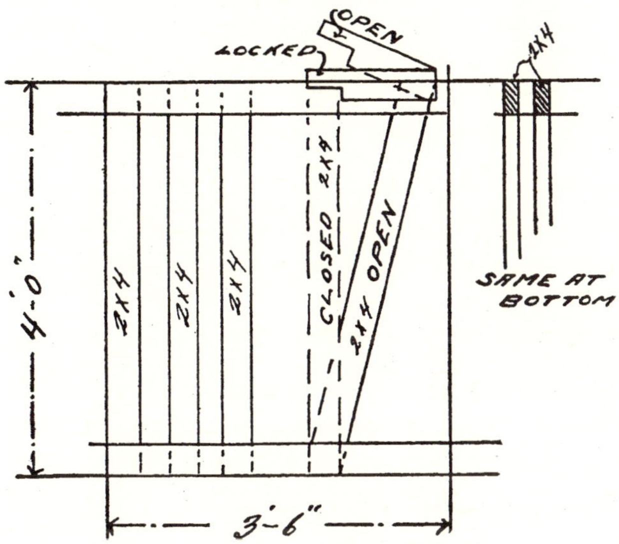

SECTION OF STANCHIONS

This way of mixing concrete is not only the best, but it is usually cheaper because it requires less cement. By actual experiments it has been proven that when the proportions are exactly right as stated the resulting concrete is not only stronger but cheaper than when a richer mixture is put together. Of course the wall under a barn like this may be poorly constructed and still support the barn for a lifetime. At the same time if you can make a wall that is harder and better than stone for the same amount of money it is much better to do so. In making the floor of course you work on a different plan because a stable floor receives a great deal of wear.

COMBINED BARN AND COVERED BARNYARD—A102

Cost of Blue Prints, $15.00

A great many dairmen like to have a covered barnyard for the cows to exercise in and some go so far as to keep the cows in this covered barnyard both night and day, just stabling them long enough to milk and feed grain and silage. In some part of the country the covered barnyard is growing in favor.

The plan, (A102), is designed for a bank sloping to the south. There is a good root cellar in the bank next to the building on the north side and the large roof surface is utilized to furnish water for the cistern. A cistern filter is placed inside the building so it won't freeze. To have nice cistern water it is best to run it through a filter.

The feed racks in the covered barnyard are made movable to facilitate driving

SIDE ELEVATION OF BARN AND YARD

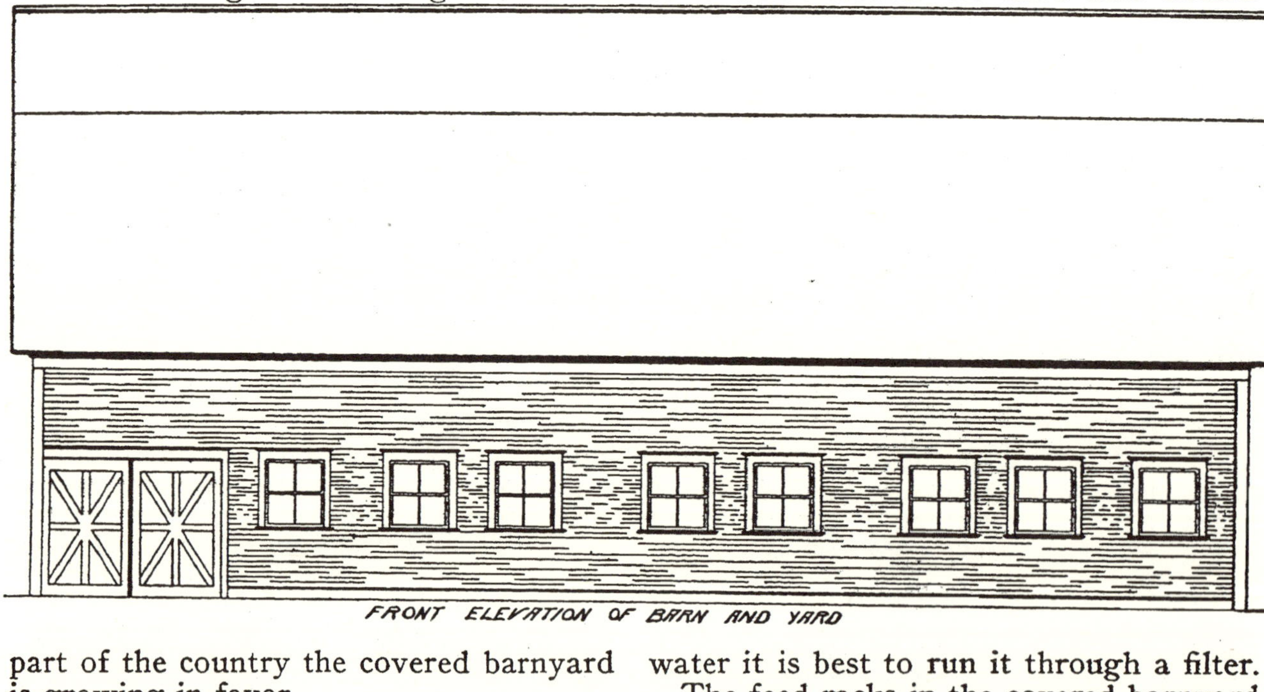

FRONT ELEVATION OF BARN AND YARD

through at cleaning time. Mild days in winter the manure spreader is brought in at one door, loaded and taken out at the other. The racks are placed in the center under the feed chutes so the roughage from the storage above may be dropped into them with as little work as possible.

A bull pen is shown in the northeast corner with the yard outside for exercise. It depends somewhat on the slope of the bank whether this is the best place to put the bull pen or not. You want the bull yard out of the way, still it should be in full view of the barn yard, both for con-

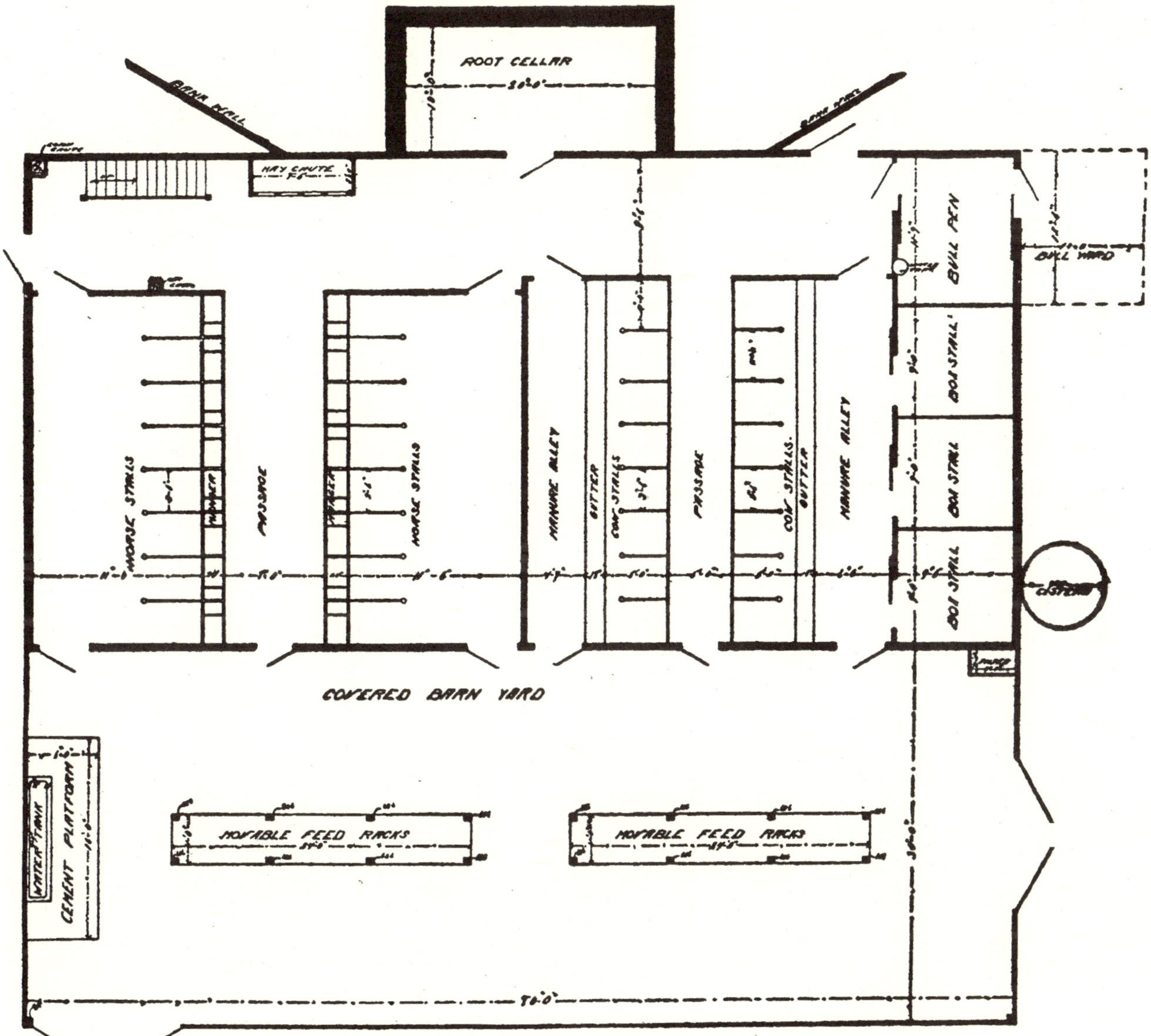

BARN AND YARD PLAN

With a cistern and a windmill the water tank is kept supplied all the time so the cows may run to it when they want to. The stable floor should be about two feet higher than the floor in the covered barnyard. This gives an eight foot ceiling for the stable and a ten foot ceiling in the yard.

venience in feeding and attending to the bull, and to keep the animal from becoming lonesome and cross. Possibly the position of the bull yard and the cistern might be reversed to the advantage of both as it would give more bank to hold the cistern and level ground for the bull.

ROUND DAIRY BARN A-205

Cost of Blue Prints, $20.00

A round barn should have a silo in the center and the silo should be a part of the barn. Round barns are not very fashionable considering the number that one may see in a day's travel, but there are many features to recommend them.

Generally speaking the greatest possible cubic space according to the amount of material may be enclosed in a round barn and the silo being in the center makes it easy to feed the stock because all the stalls may be placed within easy reach.

A cement silo is more desirable because it is durable but the first cost is greater. Wooden silos rot out in a few years' time and certain parts of them must be replaced, but this need not be taken as positive proof that the cement silo is the best under all circumstances.

Whether a cement or wooden silo is built the frame work of the barn is of light construction, and is framed into the silo. The whole structure is so braced from every direction that there is no necessity of having a heavy framework. In the first place a circular cement wall is built for the silo and another circular wall extending down below frost is built around outside of the barn. This wall reaches up about a foot above grade with the exception only of the doors to drive through and at these places the wall is widened to make a slanting bridge approach from the outside up to the level of the stable floor. These approaches are

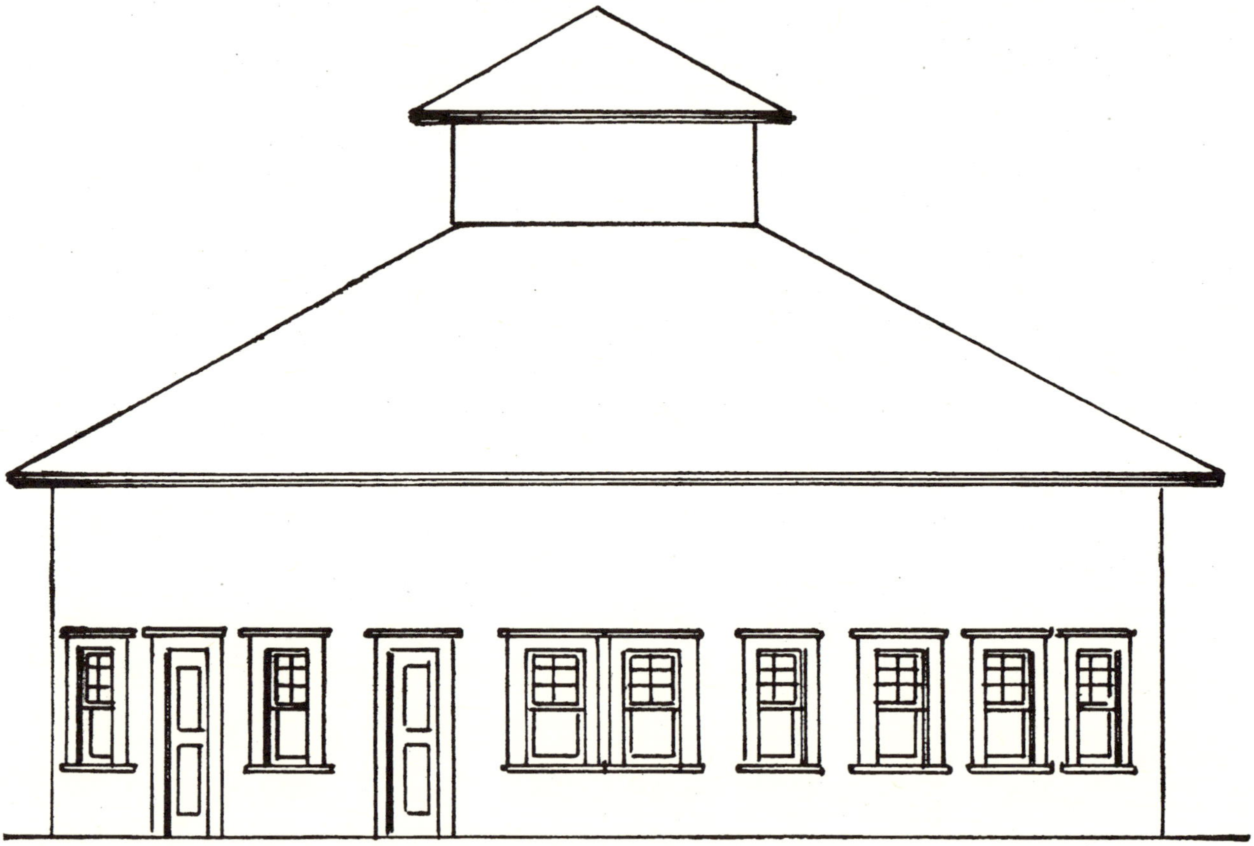

ELEVATION

reinforced with strong cement mortar to make them durable.

Circular sills may be built up on this wall or the studding may be set up on the cement and tied together both inside and holding a cask together, the only exception being at the doorways, and provision may be made here to let the door posts into a recess in the wall. Two by four or two by six studding may be used ac-

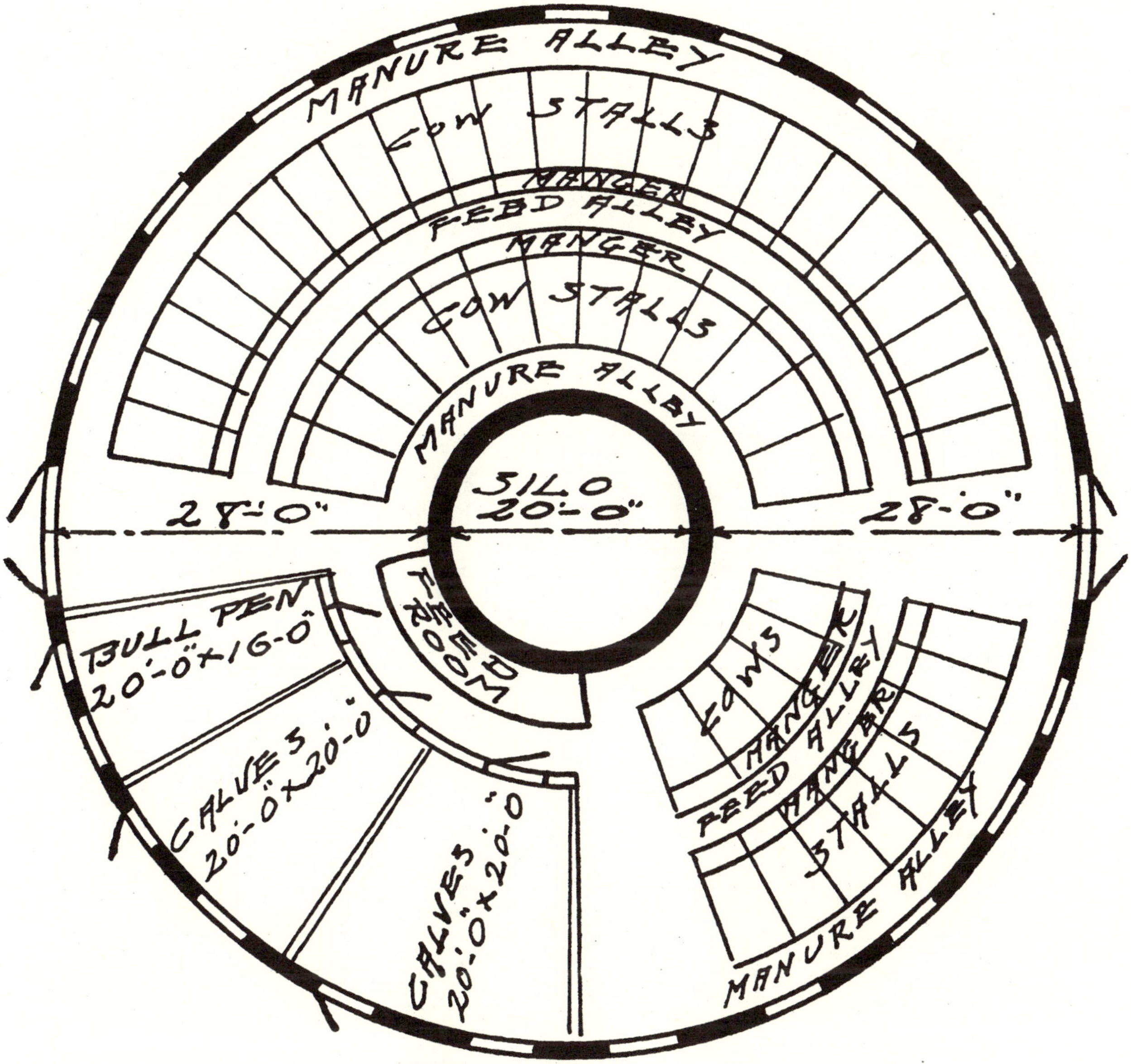

PLAN

out with an extra piece bent around and nailed firmly to each end. In fact the circular outside boarding makes the sides very solid without anything extra at the bottom because the boarding running round and round is like so many hoops cording to the distance apart and the size of the barn.

The joists all point to the center, the outer ends being spiked to the studding and they are again spiked together at the laps. The joists are suspended at intervals by girths and these girths are supported by the stall or pen partitions. It should be remembered here that the principal weight comes on these supports and they should be heavy enough to carry the load.

In a large barn like the plan shown, two by six for outside studding will be found sufficient, especially as there are so many of them. The plan shows a barn eighty feet in diameter, but these barns may be built in any size from fifty to a hundred feet with a silo in the center about twenty feet across.

The size of the silo must be determined by the number of head of stock. Silage must be fed down about two inches every day to keep it fresh and in good condition.

The roof is just as strong as the other construction with rafters all radiating from the silo to the eaves. The roof boards are put on in circles so that the roof is actually hooped, as well as the sides of the building. A cyclone might lift such a roof and roll it around the country like a cart wheel, but it would hardly break to pieces. Hay fork tracks in a round barn are suspended in circular form from a clear space left for the purpose, a little nearer the silo than the middle, because the mow grows deeper as you work towards the center.

In a round dairy barn the stalls are all built so the cows face together, which is convenient for removing the manure. A circular track for the manure carrier should extend all the way around the barn behind the cows to facilitate easy cleaning.

STORAGE BARN WITH DAIRY STABLE WING—A136

Cost of Blue Prints, $15.00

A great many dairymen object to having storage of any kind over a cow stable. There is more or less dust from the mows, and the dust is objectionable for several reasons. But it is impossible to keep cows profitably unless the rough feed and straw may be reached easily. Labor is so expensive that even the steps necessary while feeding must be counted and reckoned in the cost. If there is no storage over the

cows there must be storage near by. The silo in this plan is placed at the side about midway along the length of the stable for easy feeding. This position also makes it easy to get the green cut stalks into the silo at filling time.

SIDE ELEVATION

Not much corn is fed to dairy cows, but the crib is not far away from either cows or horses. A dairy room ten by seventeen feet is built in the corner next to the storage barn easily accessible to the stable. It would not be necessary to have a door at the outer end of the feed alley, but it is very handy in the summer time if green feed is fed to the cows in the stable.

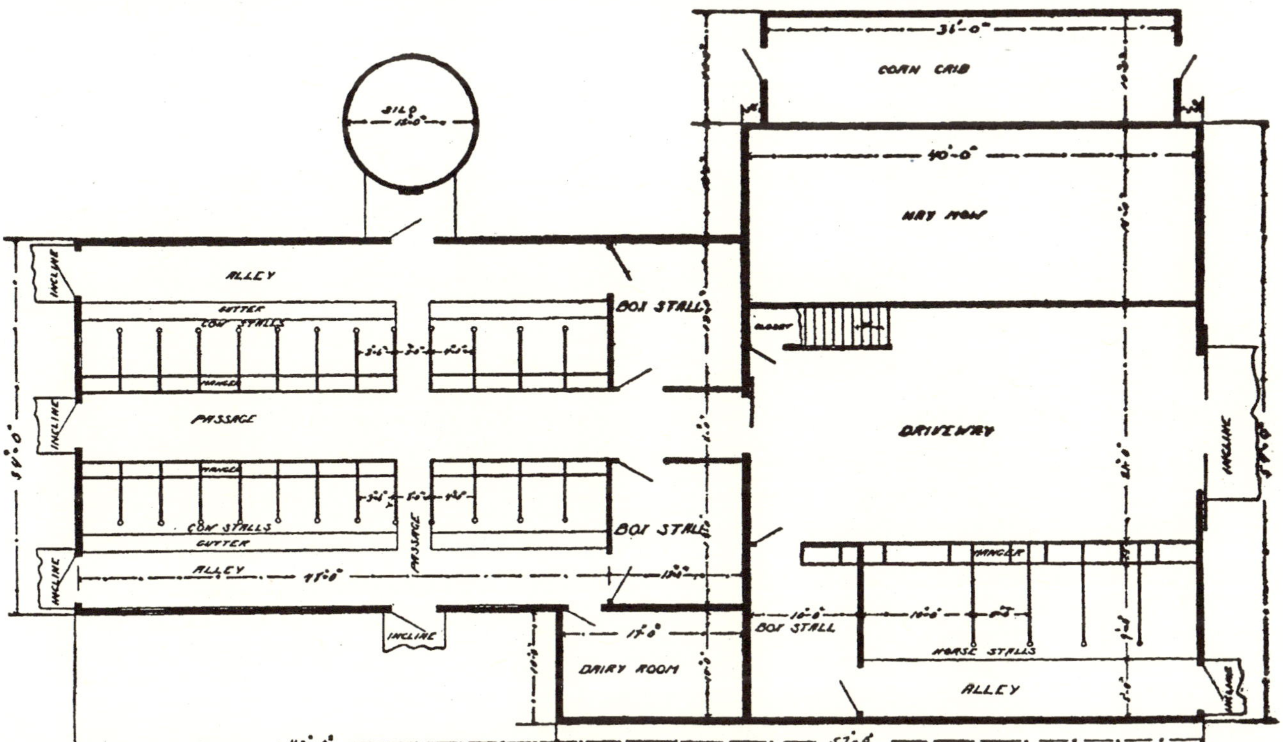

FLOOR PLAN OF CATTLE BARN

In the storage barn the hay mow reaches from the ground to the roof. For comfort in cold weather it is better to board up the side of this mow to the floor over the driveway. It is necessary to have an opening through the floor over the driveway to use a hay fork. The opening may be boarded around and used as a chute to pitch hay down through for feeding, but such details must depend to a great extent upon the kind of farming carried on and the other buildings on the farm.

MODEL COW BARN—A158

Cost of Blue Prints, $15.00

The size of this cow stable is thirty-eight feet six by one hundred and forty-two feet and it has a capacity for housing fifty-two cows. It was designed very carefully to provide every comfort for a herd of thoroughbred Guernseys.

The entire floor is made of concrete, including manger and manure drains which carry the liquid manure back to the manure pits. They are also connected with the sewer drain so that the wash water from flooding the floors can be carried away to a safe distance.

The mangers are also connected with the sewer so that the cows may be watered in the manger and the surplus water immediately drawn off.

A space of two feet high between the studding of the outer walls is filled in with concrete and troweled smooth with a curve at the floor line to leave no chance for the collection of filth to favor the breeding of disease germs. Gas piping is used for stalls set firmly in the cement. Each stall is finished with individual wrought iron hay racks made to swing up.

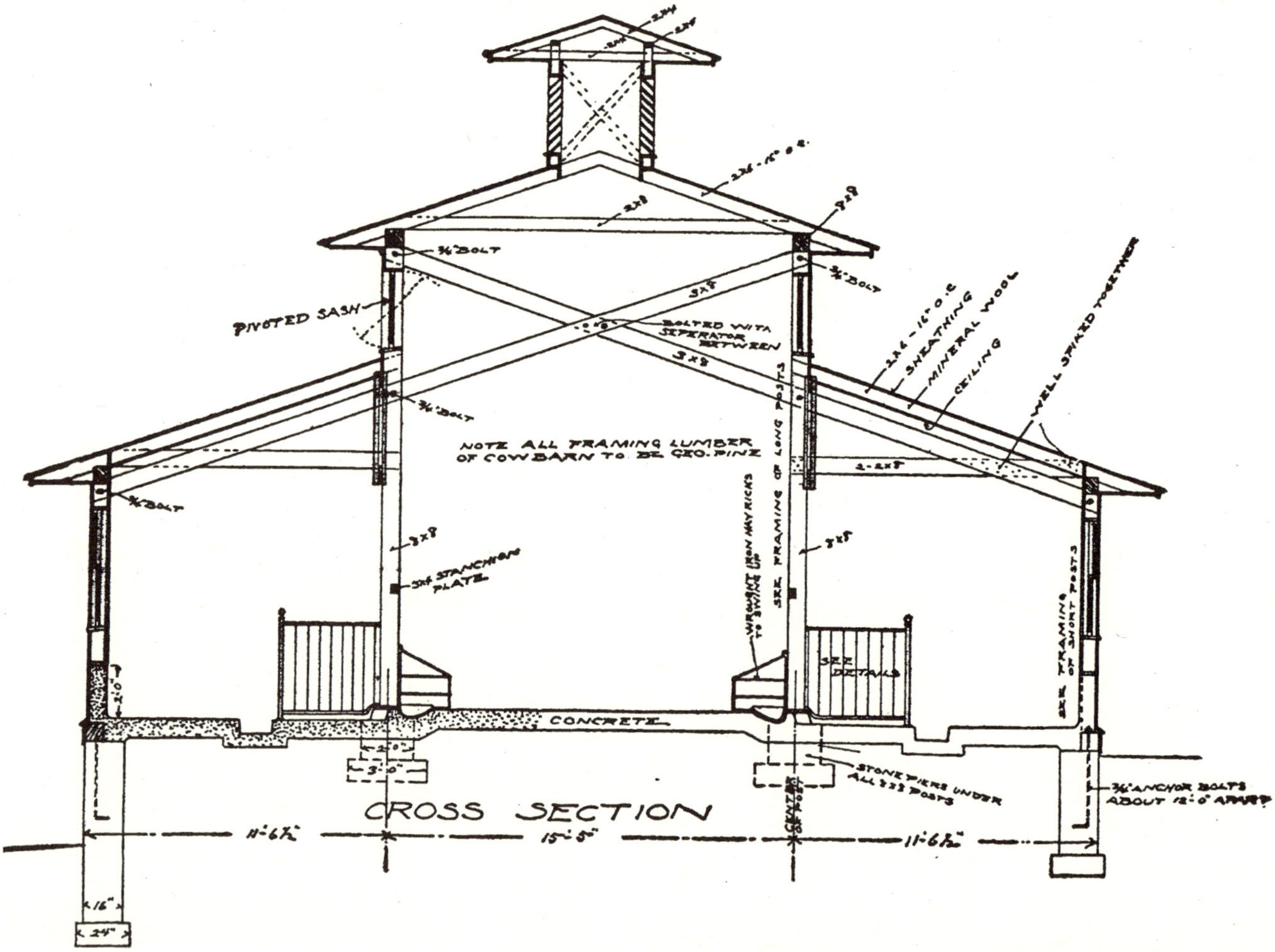

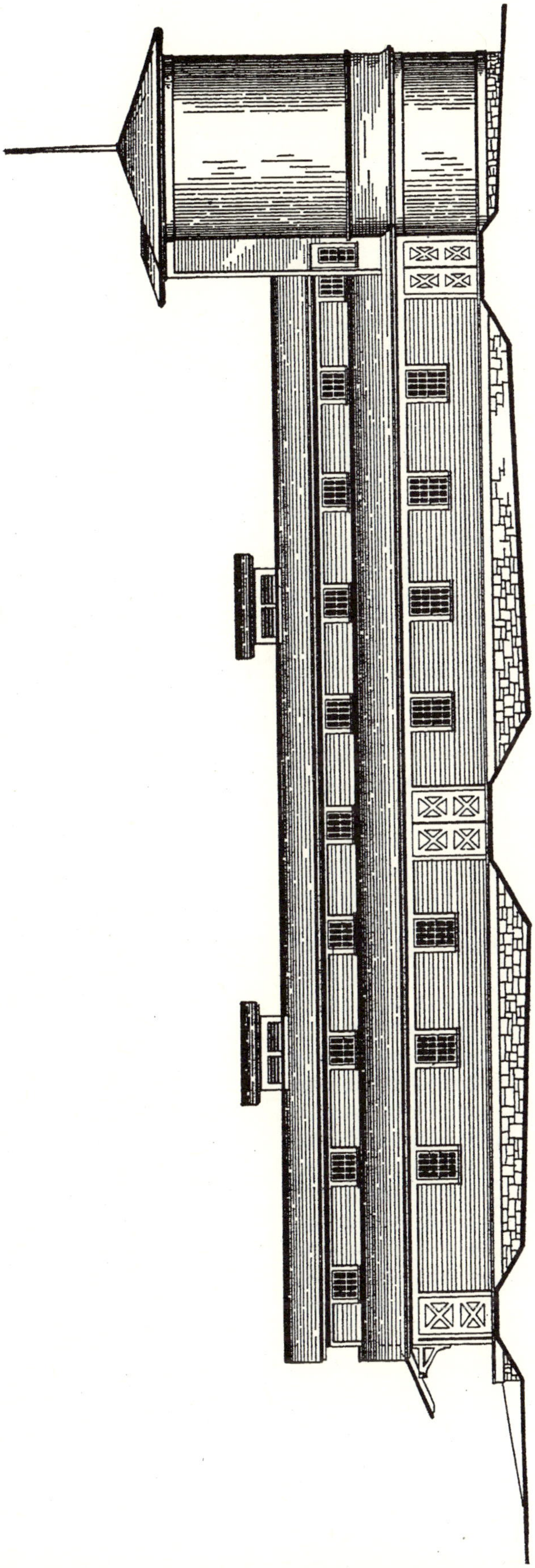

SIDE ELEVATION

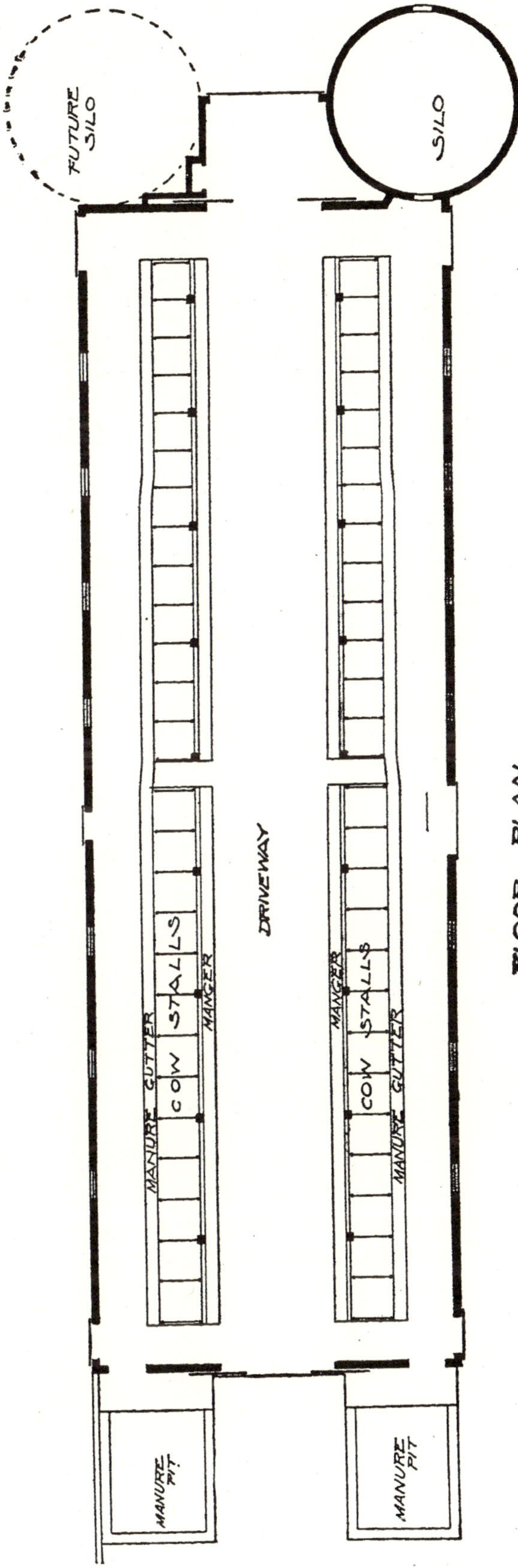

FLOOR PLAN

There is a cement top to the concrete floor which is finished rough enough to prevent slipping and to hold the bedding. Cows in this stable face towards the center and the center aisle is wide enough to drive through with a wagon and hay-rack for hauling loose hay and fodder. The silos are located at the end, the silage being loaded into cars and wheeled through the feed alley to the mangers.

Light and ventilation were main features in the construction of this stable. Careful calculations were made to secure plenty of fresh air for each animal as the sanitary conditions with such a valuable herd of animals is an important feature.

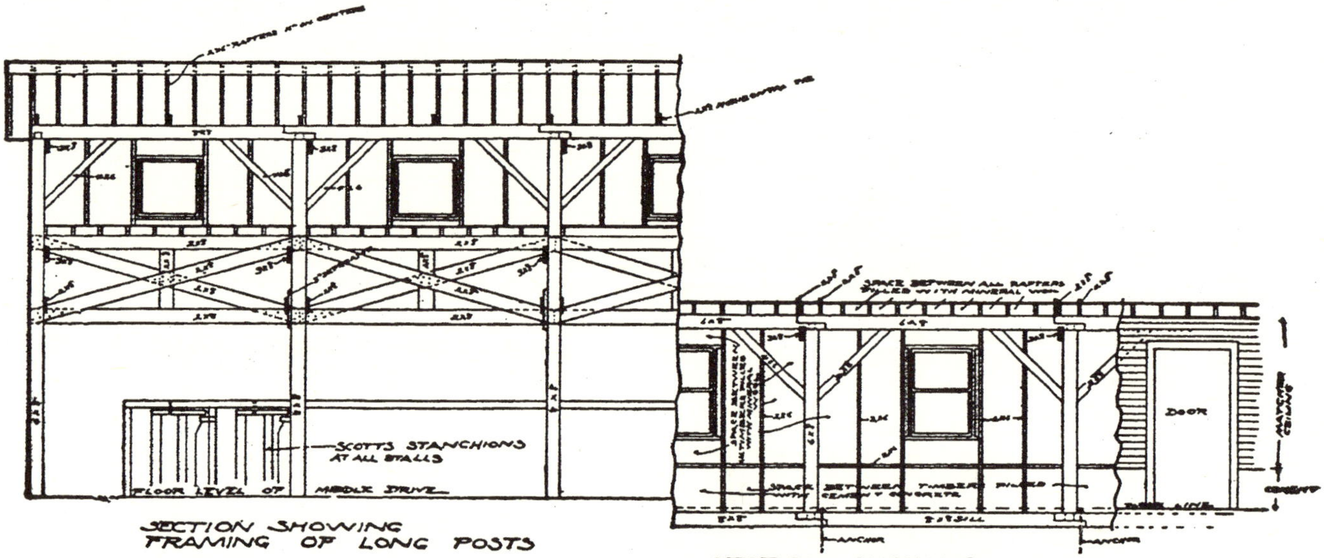

SECTION SHOWING FRAMING OF LONG POSTS

SECTION SHOWING FRAMING OF SHORT POSTS.

BARN FOR DAIRY COWS—A162

Cost of Blue Prints, $10.00

This cow barn is forty feet wide by eighty-one feet long and will accommodate twenty-four cows. There is a feed room, wash room for washing utensils and an office. Along one side a silo is placed near the mixing room and convenient to the feed alleys which in this stable are at the sides.

The manure gutters and floor for cleaning is in the center so that in this stable the cows face outward. This arrangement makes it easier to remove the manure and the plan is liked by some dairymen.

The balloon roof construction makes it possible to store a great deal of feed over head. It leaves a clear space for the horse fork which works freely from one end of the building to the other. Roofs like this are comparatively new. The first ones built were not strong enough to stand heavy winds and some of them blew down, but there has been no such trouble recently. If properly braced each side forms a truss and the two trusses meet together at the peak.

There are hay chutes at the sides for putting down hay and bedding and there is a stairway at the side of the office for convenience in getting up and down.

To help out in feeding time there should be a silage carrier to run from the silo down the different alleys to distribute the feed. If a farmer wants to know the number of miles traveled about the stable it is only necessary to figure the number of trips and steps taken each feeding time, then multiply this by the number of feeds during the winter. If every dairyman would do this the location of some silos would be changed. The amount of travel

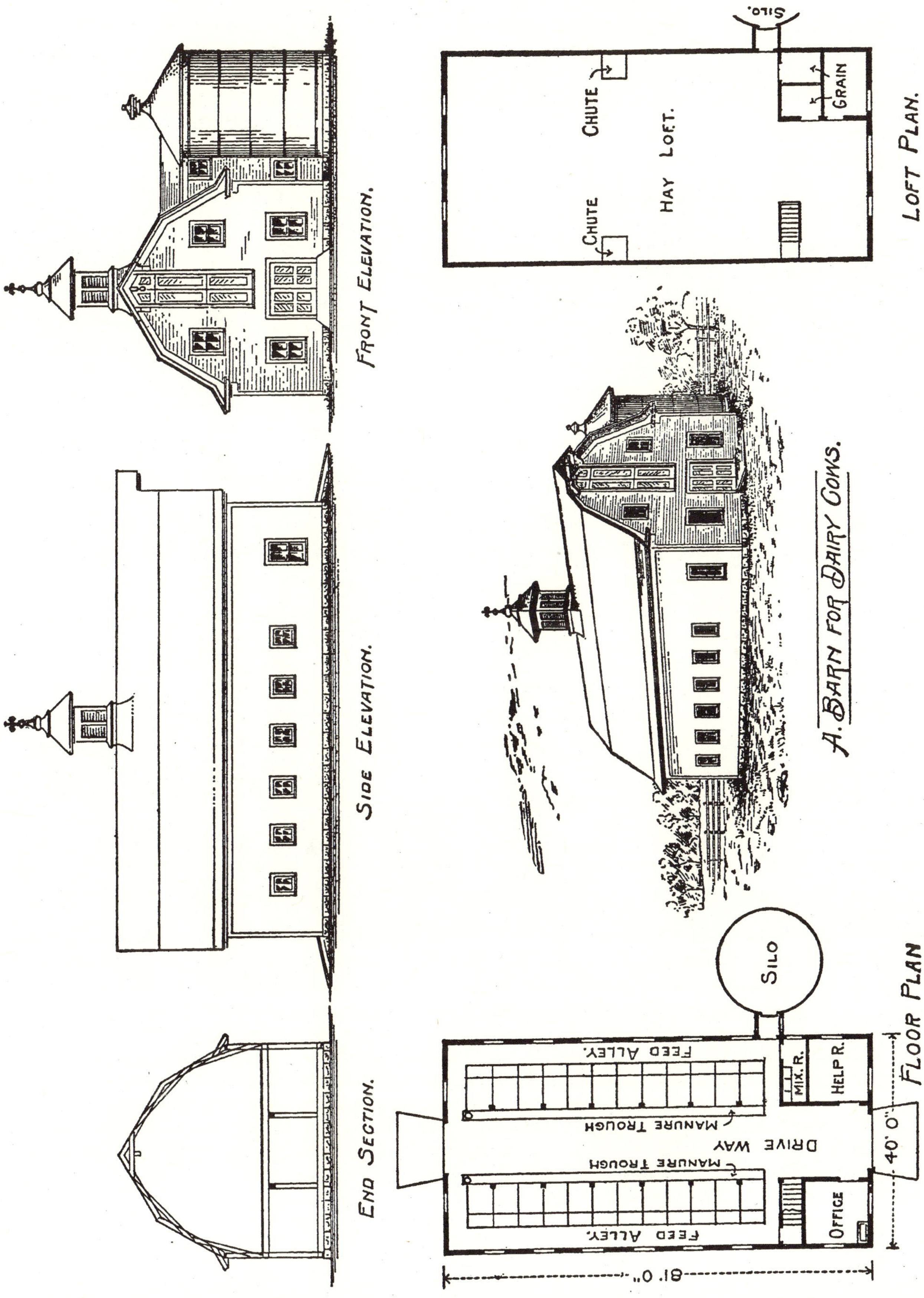

A BARN FOR DAIRY COWS.

will surprise those who have never thought about it. This is one reason for placing the silo at the side.

The manure alley in the center is wide enough to drive the manure spreader right through, loading it in the meantime so it is not necessary to have a pile of manure outside of the stable. Manure is worth a great deal more when it is drawn immediately from the stable to the field. This barn looks well and it is a good practical barn.

STABLE FOR TWENTY-FOUR COWS—A101

Cost of Blue Prints, $15.00

This plan provides stabling complete for twenty-four cows with calf pen, bull pen, two box stalls, a feed room and a wash room. This plan offers the advantage of a wide driveway through the center feed alley which is a great advantage in the summer time when green feed is used for soiling purposes and hauled directly on hay-racks from the fields to the cows in the stable.

SIDE ELEVATION OF DAIRY BARN

There is an advantage in having a bull pen arranged in this manner. The door at the corner opens into the yard for exercise and the pen inside is made of one and one-half inch gas pipe pickets placed five inches apart from centers. This gives about three inches in the clear between the pickets. The object in this is to let the bull see everything that is going on in the stable. It makes a bull much more contented and he is less liable to become cross. A bull needs company just as much as any other animal. A great deal of trouble has come from shutting bulls up in tight pens where they become lonesome and morose. Box stalls are boarded to the ceiling and made as warm and as comfortable as possible.

The width of this stable is thirty-six feet, rather wider than usual but it allows ample room for the driveway in the center and a good passageway behind the cows besides giving room enough to place the feed room, box stalls and other pens on opposite sides of the driveway in one end of the stable. The length of the building is eighty-four feet, but of course it could be extended if more room is desired without altering the width or the general plan.

Placing the silo near the middle of the building saves carrying the silage more than fifty feet which is a great saving of steps at feeding time. One great advantage with this stable is the number of windows. The windows extend from the ceiling to within three feet of the floor which is a great advantage in admitting sunshine. The manger in this stable is placed two inches above the floor. It is two feet wide and six inches deep and the bottom is slightly rounded. Three feet six inches are allowed for the width of the stalls with a standing floor four feet ten inches. Of course both the length and width of the

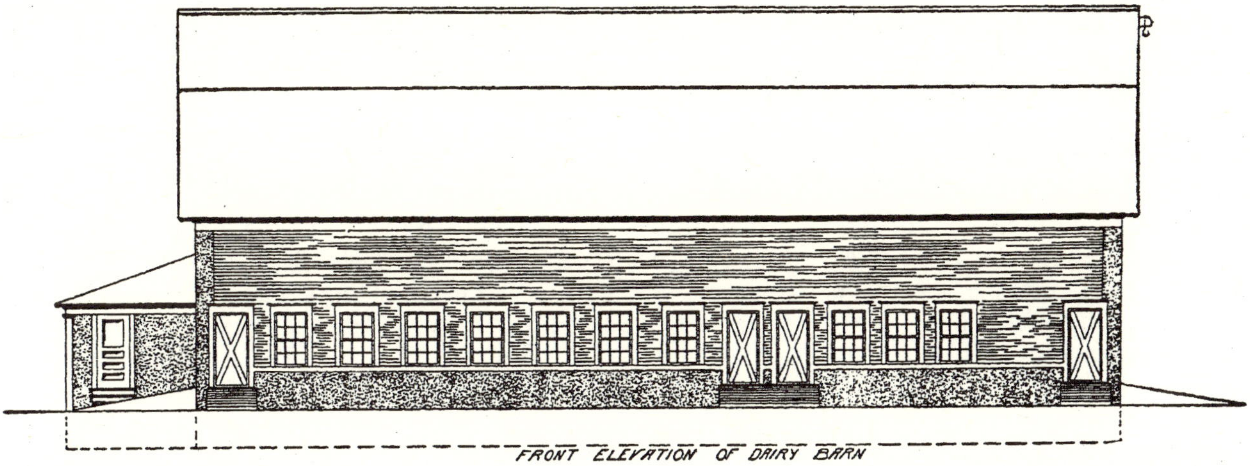

FRONT ELEVATION OF DAIRY BARN

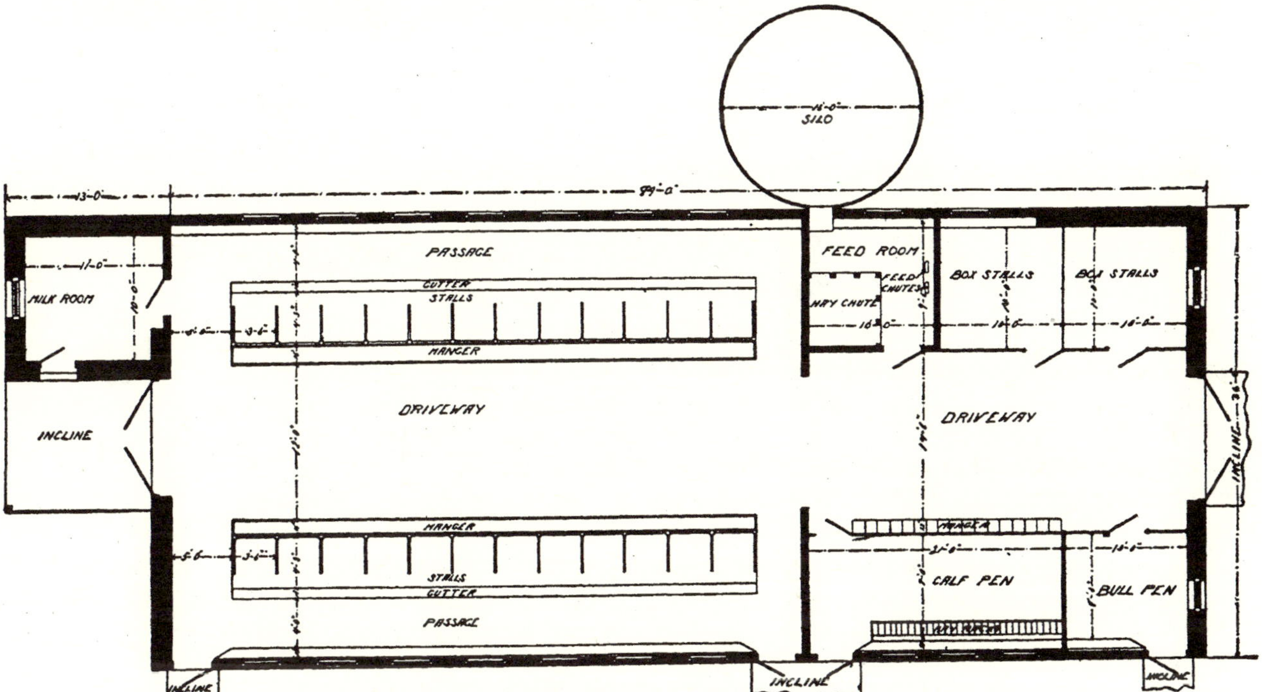

FLOOR PLAN OF MODERN DAIRY BARN

stalls should be made to fit the cows. For an extra large Holstein a four foot stall with a five foot length may not be too much but four feet ten inches by three and one-half feet is big enough for most cows and it is too much for some. A perfect cow stall has never been invented. If some dairymen wishes to be honored by posterity he should get busy and invent a cow stall that will be thoroughly satisfactory under all circumstances.

The calf pen in this plan meets the views of the best dairymen who have examined it. It is twenty-one by eleven feet with a manger in front for grain feeding and a hay-rack along the back wall. Individual stanchions are provided for use when feeding the calves grain or milk. It would be difficult to devise a better arrangement for calves and we all know that the calves of this year are the cows of two years hence and the value of the cow depends on the quality of the calf and the feed and care given it.

A silo for twenty-eight cows should hold about one hundred and thirty tons. This amount will rather more than feed the cows during the winter but it is a good plan to have a little silage left over to help out the green feed in summer time. A silo sixteen feet in diameter and thirty-two feet high is very satisfactory.

The milk room is not exactly separate but it is built on the front and there are two spring doors to shut out the odors of the stable. This building provides for storage over the stable with a feed chute in one corner of the feed room. There is a large door between this feed room and the alley for the purpose of preventing dust from flying out into the stable. This feed chute is large enough so that hay, straw or any roughage may be dropped into it from above in sufficient quantity at one time. The door may then be opened and the stuff forked out. There is also a small door opening from the chute into the feed room. This is for the purpose of mixing together feed with chopped stuff in case the owner puts a cutting box overhead.

Because of the storage room above, the upper floor is made double thickness with two thicknesses of paper between, matched flooring is used and the first course nailed to the joists in the usual way, only that the dressed side is placed down. The two thicknesses of paper are then put on and the other floor laid over it and nailed over the joists, the workmen being guided by chalk lines on the paper.

COW BARN FOR FORTY COWS—A159

Cost of Blue Prints, $15.00

A cow barn for the accommodation of forty cows having a feed alley of sufficient width to accommodate a wagon with a load of soiling feeds is shown in this plan. This is the quickest and cheapest way of distributing feeds to the mangers along both sides of the feed alley.

The mangers as well as the whole floor surface are built of concrete with the mangers elevated only three inches above the floor level. As cows naturally feed from the ground it is only right that the mangers should be very low down. The side of the manger nearest the cow is made almost perpendicular to prevent feed from working over amongst the bedding. But the feed alley floor is elevated and that side of the manger is rounded up to it which makes it easy to keep the feed in the mangers and easy to kick it back when the cows shove it out, as they do while feeding.

A water faucet is placed at each end of the manger for the purpose of watering the cows. For disposing of the water left in the manger a drain in the center with an overflow is provided. The middle posts extend from the back of the mangers and run to the roof and these are spaced to allow three stanchions between the posts.

A gutter sixteen inches wide and from five to eight inches deep is run diagonally behind the cows, starting at five feet four from the mangers at one end and finishing

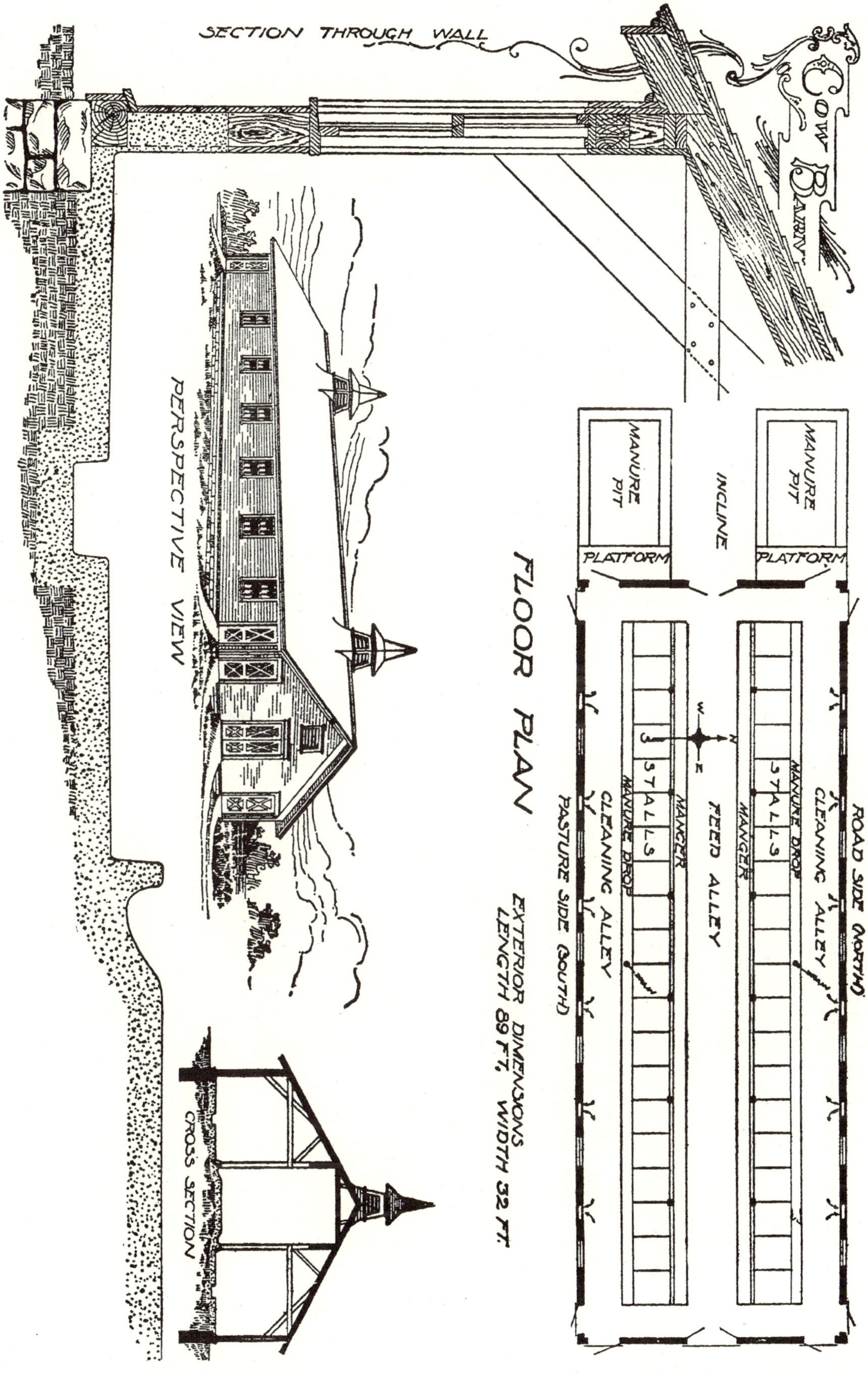

COW BARN
SECTION THROUGH WALL
PERSPECTIVE VIEW
CROSS SECTION
FLOOR PLAN
MANURE PIT
INCLINE
MANURE PIT
PLATFORM
PLATFORM
ROAD SIDE (NORTH)
CLEANING ALLEY
MANURE DROP
STALLS
MANGER
FEED ALLEY
MANGER
STALLS
MANURE DROP
CLEANING ALLEY
PASTURE SIDE (SOUTH)
EXTERIOR DIMENSIONS
LENGTH 89 FT. WIDTH 32 FT.

up at the other end five feet ten, thus making different length stalls to accommodate longer or shorter cows.

The floor of the stalls is given a slight slope from the manger back to the gutter and the surface of the floor is left rough to prevent the cows from slipping and to hold the bedding in place. There is sufficient room back of the alley to run a truck or wheelbarrow to facilitate cleaning out the manure. The liquids of course run to the lowest point in the center of the gutters where they are connected with a bell trap drain, whence they are carried to a catch-basin directly opposite the drain outside of the building. From this catch-basin the liquids are pumped into the distributing manure cart.

All side walls are filled in solid between the timbers with cement concrete to a height of two feet above the floor and then finished with smooth cement plaster which makes a perfectly sanitary finish and permits the entire barn floor to be washed with a hose and flooded with water without injuring any woodwork.

Warmth and ventilation are secured by fitting the size of stable to the number of animals and there are windows enough to admit abundant sunshine which is nature's best disinfectant. Ventilators and fresh air shafts in the walls supply a continuous stream of fresh air which can be controlled by slides. The foul air enters the shafts near the floor and rises in the walls to the triangular vent duct under the ridge of the roof and from this duct the air is exhausted through the slat ventilator towers. About 1,800 cubic feet of air space is provided for each animal.

BARN AT MT. CARMEL, ILL.—A195

The barn shown is on a one hundred and twenty acre farm near Mt. Carmel, Ill. It is simple in outline, but commodious and very serviceable.

The foundation is of concrete which extends 7 ft. deep under east end of driveway forming a cellar 7x12x20 ft. in which are stored fruits, vegetables and perishable

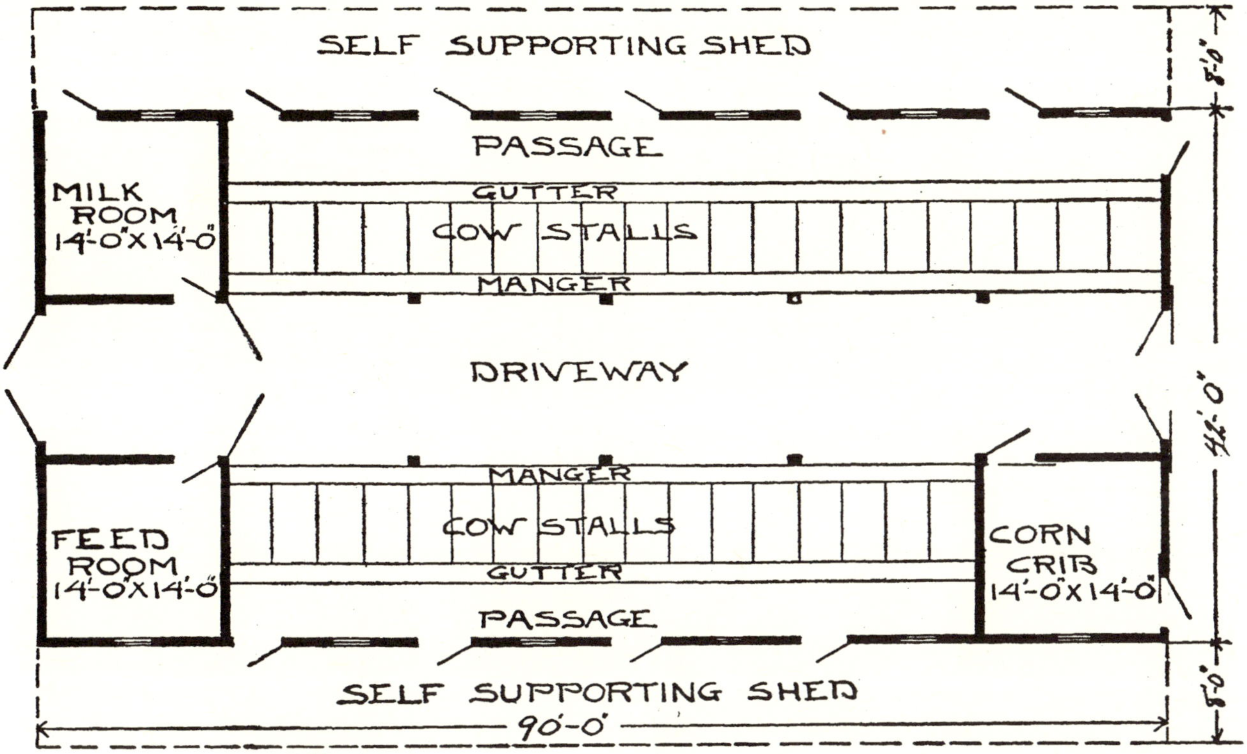

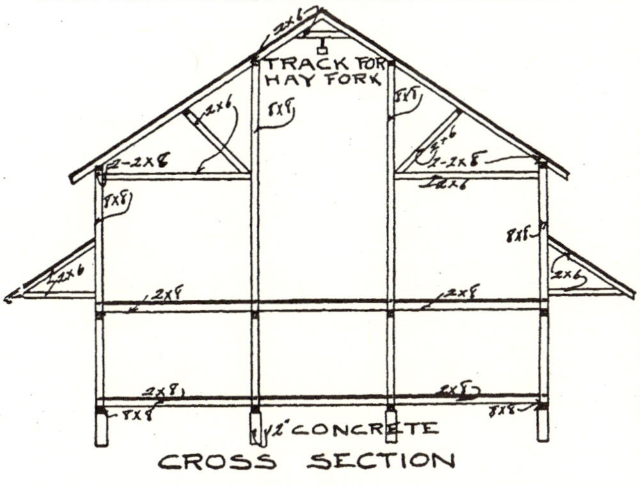

foods and feeds for both family and live stock. With rope-and-pulley device a barrel of apples or a bag of potatoes may be lowered in or drawn up from this cellar. Elevation and plans give general features of the structure, showing that the barn is intended to stable thirty-eight cows and there is provision for a small corn-crib, feed room and milk room. The idea is that later when the business grows to demand it these rooms will be removed to an outside building, or separate buildings, and the whole floor space of this barn given up to the stable proper.

TWENTY-FOUR COW STABLE—A210

Cost of Blue Prints, $15.00

It will be noticed that about one-third of the ground space of this dairy stable building is taken up with creamery, delivery room, feed room and work shop. The reason for this is that there must be no tinkering work done in the creamery or delivery room. These two compartments must be kept as pure and clean as possible, or the brand of milk required by the boards of health cannot be manufactured. Unless there is a shop for tinkering fitted with a desk and a cupboard to hang extra clothing these things will accumulate in the creamery and delivery room. Then when the inspector comes around, the business

END ELEVATION

of that farmer is classed as second or third rate according to conditions as he finds them.

being taken to use only the best materials and to put them together quickly so the job will harden at one time and set together in one great stone. Above the concrete wall the structure is of wood, but the

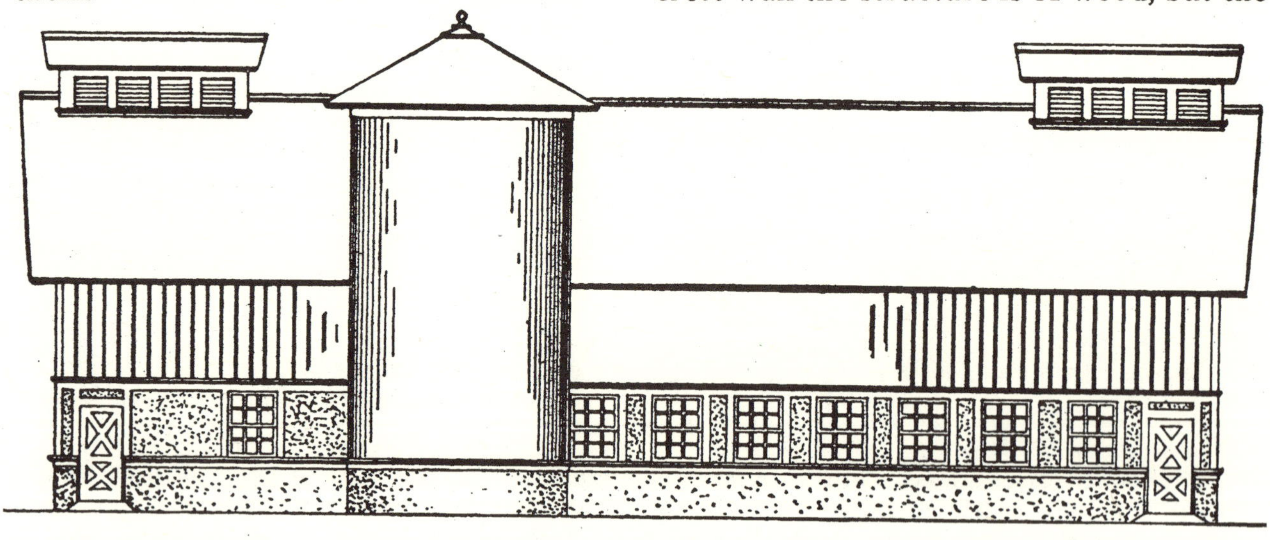

SIDE ELEVATION

The wall and floor of this dairy barn is built all in one piece of cement, great care lower story is plastered outside with cement. All windows are carefully fitted to keep out the cold in winter and they are supposed to be kept clean at all times to let in plenty of light.

There is a difference of opinion among dairy farmers in regard to the center driveway between the mangers, some claiming that this space is greater than necessary and that it adds unnecessarily to the width and consequently to the cost of the building, but those dairymen who feed their cows in the stable in the summer time as well as in winter like to drive through with a hay-rack loaded with green feed brought directly from the fields, because they can feed the cows so quickly and with the least possible labor.

The feeding may be done in the same way in winter by bringing roughage from the storage barn, as the same driveway will accommodate a large feed box on wheels which may be pushed back and forth to the silo and feed mixing room.

Especial pains is taken with the partition between the cow stable proper and the creamery room. It is made solid and the door is carefully fitted and supplied with a spring to keep it shut in order to protect the milk from stable odors. Another important feature is to front the stable towards the prevailing winds so the draft will be from the creamery end out through the stable. Some dairymen make the mistake of creating a draft the wrong way and it makes a lot of difference in the scientific production of clean milk.

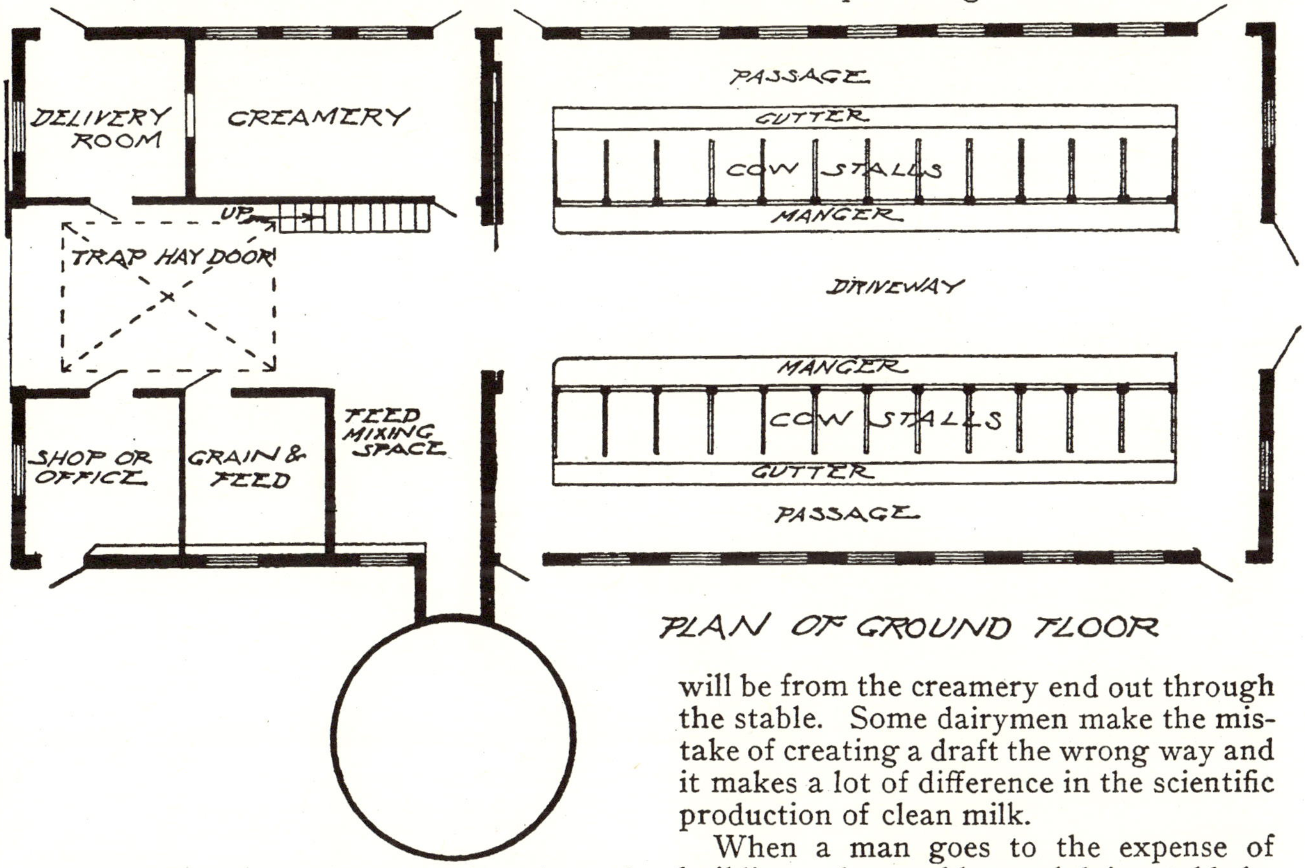

PLAN OF GROUND FLOOR

When a man goes to the expense of building a thoroughly good dairy stable he expects to manufacture high grade milk and to secure from one cent to five cents per quart bottle more than those farmers who work along in the old fashioned way. But unless all these details are considered and carefully worked out he will have difficulty in getting his price.

PRACTICAL COW BARN—A208

Cost of Blue Prints, $15.00

This cow barn is designed for a cold climate and a special effort was made to protect the stock from the cold, and at the same time give them proper ventilation and a continuous supply of fresh air by means of air ducts built in the walls, which receive the air near the ground level and the floor and carry it to the ventilators on the roof, which are controlled by a cord, regulating the flow of air as desired.

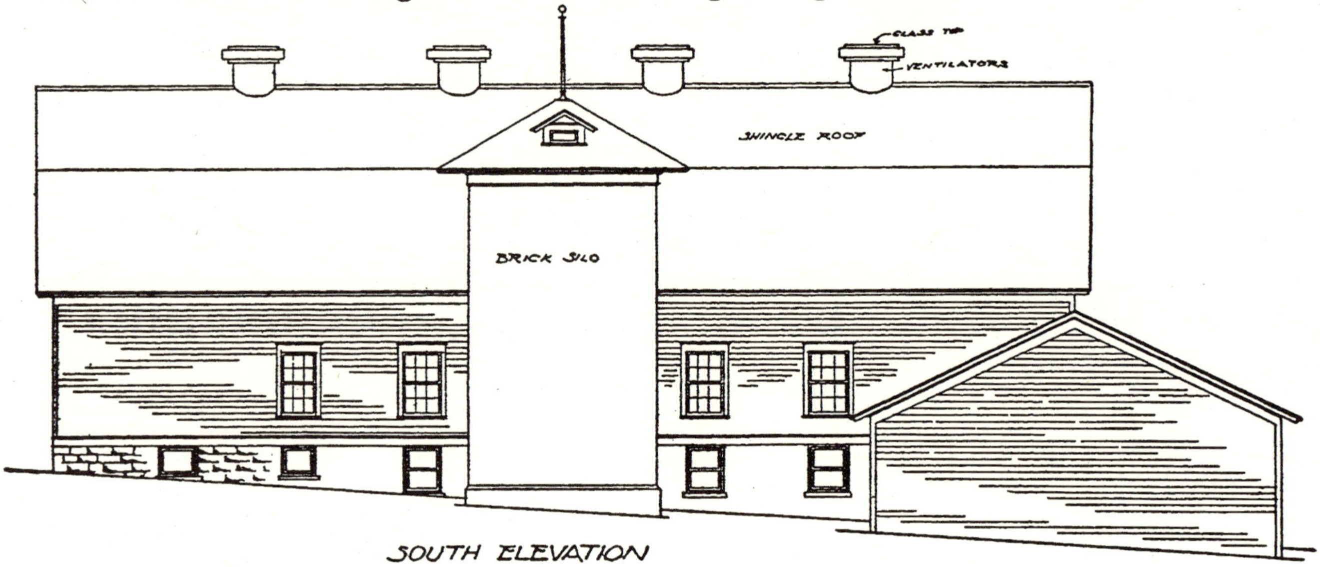

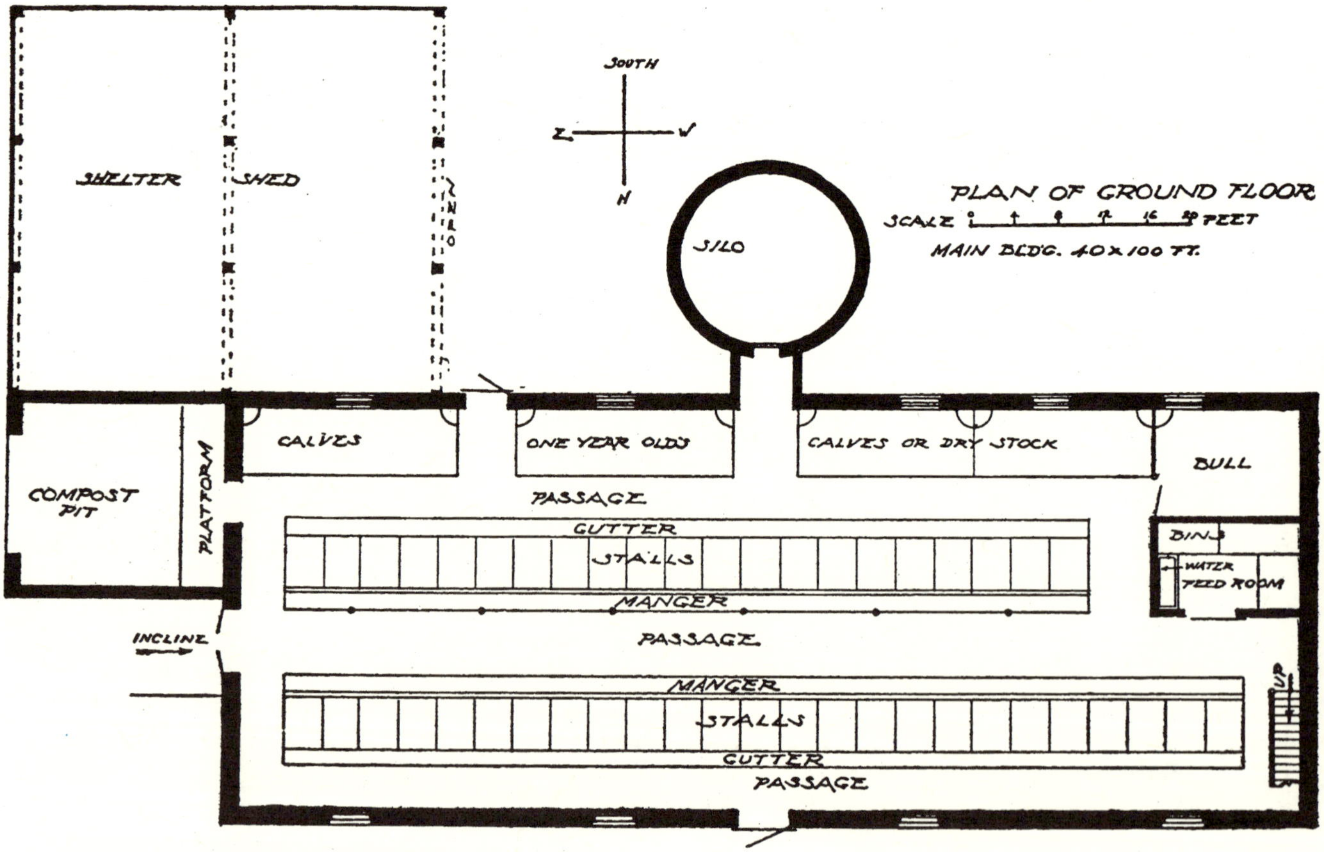

conduct it to the inside of the barn where it enters the stock room near the ceiling. Other ducts exhaust all the foul air from

The barn is located on the slope of a hill so that the hay can be hauled directly into the upper floor, and the walls of the stock room are built of rubble stone 18 inches thick; this, together with the hay above, makes a warm stock room.

The silo is located at the center of the south side, where it is convenient for feeding and also protected from the north winds. The shelter shed is also located on the south and at right angles to the main barn, so that the stock is well protected when out of the barn in severe weather. As will be seen from the drawing, this barn is 40 by 100 feet, and contains stalls for forty-six milch cows, besides loose stalls for calves, dry stock, bull, etc. At the west end is a feed room with bins connected by spouts to larger bins on the floor above; also stairway to the upper floor, and on the east end is a manure pit covered by an extension of the shelter shed roof.

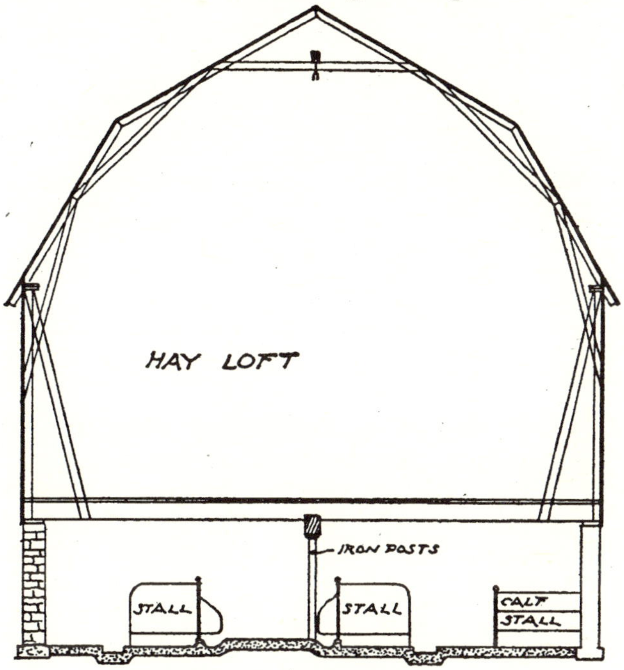

CROSS SECTION

The cross section clearly shows the general arrangement of stalls, mangers, gutters, etc., all constructed out of cement laid on solid ground. The stall partitions are built up out of wrought iron bars and pipes, leaving nothing to get out of order or decay. The wood superstructure is constructed out of plank, and the roof is self-supporting, without posts or purlins, by each set of rafters braced, forming a continuous arch from one sill to the other.

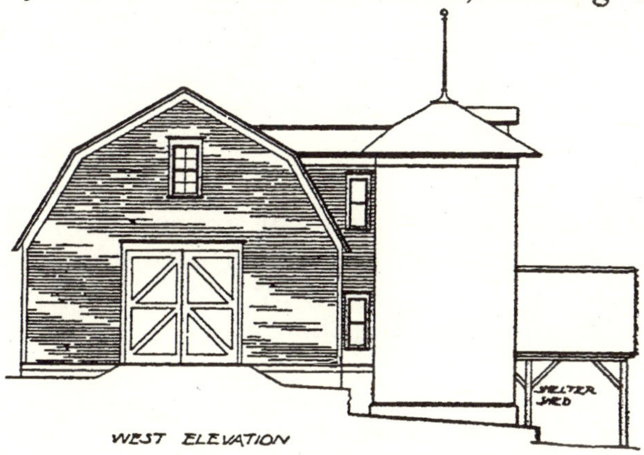

This roof gives an enormous capacity to the hay room and is well braced against sagging and wind pressure.

The exterior of the barn is sided with matched siding and the roof is of shingles, making a very durable and good looking building, and at the same time a barn that can be built within a reasonable figure.

FORTY COW BARN—A209

Cost of Blue Prints, $25.00

A large modern dairy farm building of exceptional completeness and convenience of arrangement is presented herewith. In the one building are grouped the accommodations for the three branches of dairy farming: There are, first, sanitary stabling for forty milch cows, twelve dry cows, two bulls and numerous calves; second, improved storage capacity, accurately figured, for ensilage, grain, roots, dry fodder and bedding sufficient for that number of cattle; and, third, a well-equipped milk cooling and shipping department.

For a barn of such large capacity the arrangement in this case is very good. The general form of the building is that of a cross. The stanchions are arranged in a double row on the ground floor of the long cross member; the feed storage section, extending back at right angles to this, joins it at the middle. There, two large silos are located with the feed mixing floor between—thus having a very central location. The grain bins are next the silos, extending 28 feet from the ground floor to the plate and provided with a continuous cup conveyor, operated by a small electric motor, for elevating the grain.

filled from the outside through inclined chutes. In that way the reinforced concrete floor was not weakened by trap door openings through it.

This design provides for the rain water from the roof to be conducted to two buried cisterns, from whence it is pumped to a large tank overhead, as needed.

In the front of the building, completely separated from the barn and stable, are

FRONT ELEVATION

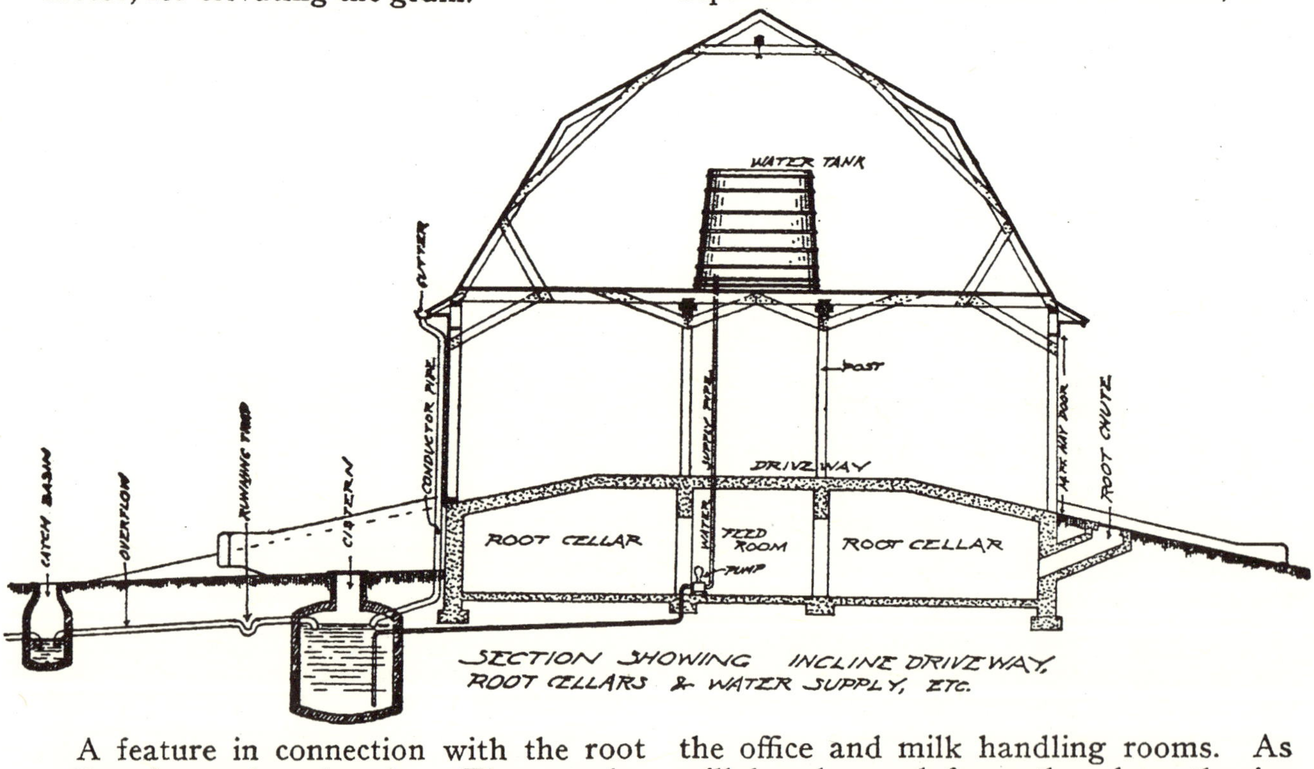

SECTION SHOWING INCLINE DRIVEWAY, ROOT CELLARS & WATER SUPPLY, ETC.

A feature in connection with the root cellars is worthy of notice. They are located on the ground floor under the "barn floor" or elevated driveway. They are

the office and milk handling rooms. As will be observed from the plans the ice house is very conveniently located to the cooling room. A detail drawing of the

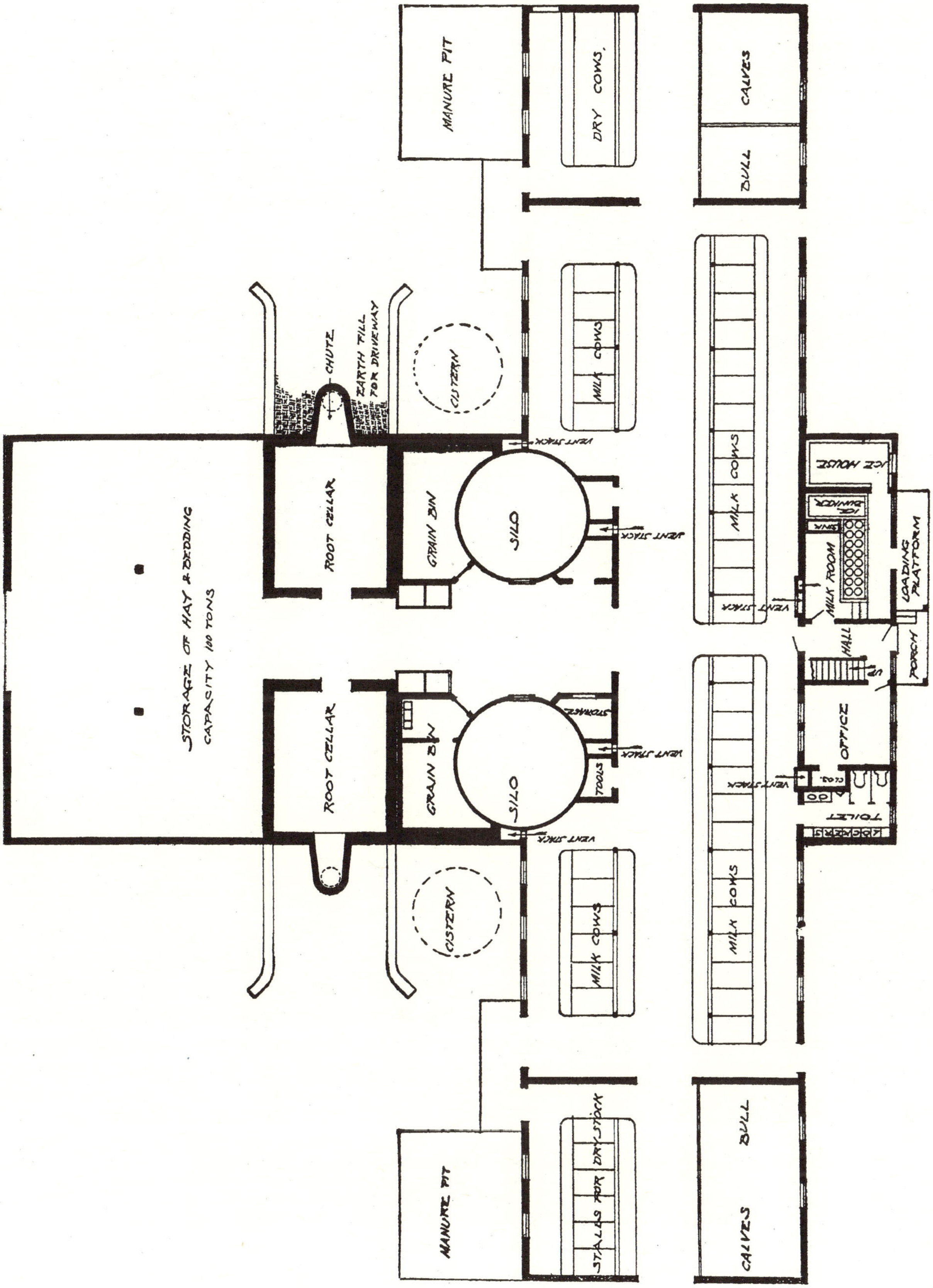
MANURE PIT
DRY COWS
CALVES
BULL
MILK COWS
CISTERN
CHUTE
EARTH FILL FOR DRIVEWAY
ROOT CELLAR
GRAIN BIN
SILO
VENT STACK
ICE HOUSE
SINK
MILK ROOM
LOADING PLATFORM
HALL
UP
PORCH
OFFICE
TOILET
TOOLS
STORAGE OF HAY & BEDDING
CAPACITY 100 TONS
STALLS FOR DRY STOCK

milk cooler is given, showing the ice bunker and compartment for the milk cans in

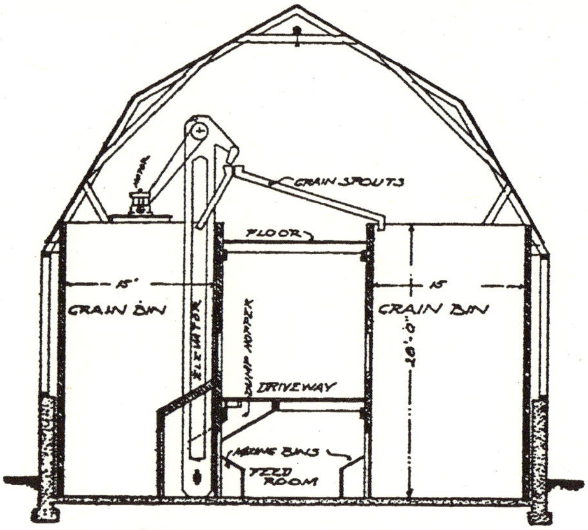

SECTION SHOWING GRAIN BINS. GRAIN ELEVATOR, MIXING BINS, ETC.

cross section. All the walls and covers of this chest are made very heavy, built up in

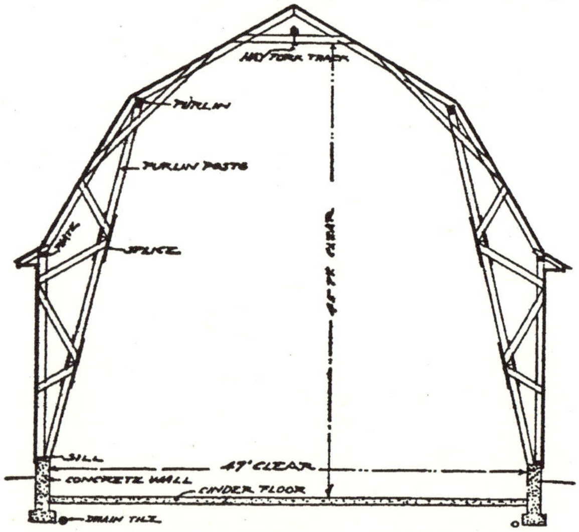

SECTION THROUGH HAY ROOM

layers of insulating material to keep out the heat. It might be interesting to experiment with the calcium chloride cooling medium in connection with a refrigerator made like this.

Calcium chloride is a substance which increases the capacity of brine to maintain

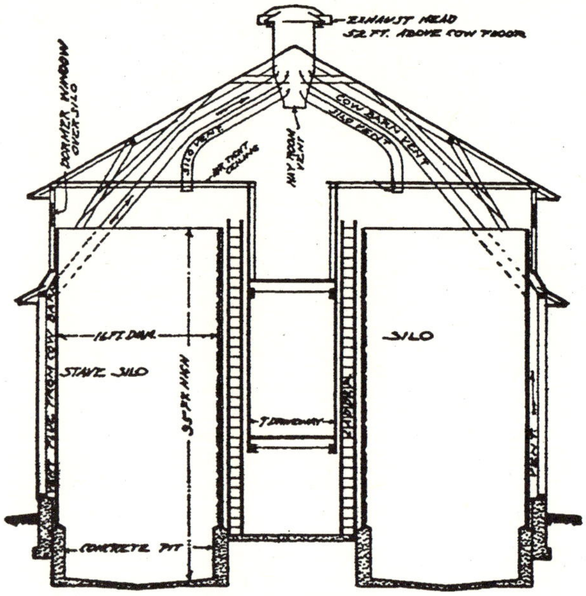

SECTION SHOWING SILOS AND VENTILATION FLUES

a low temperature. The calcium chloride brine circulates in the pipes and is cooled

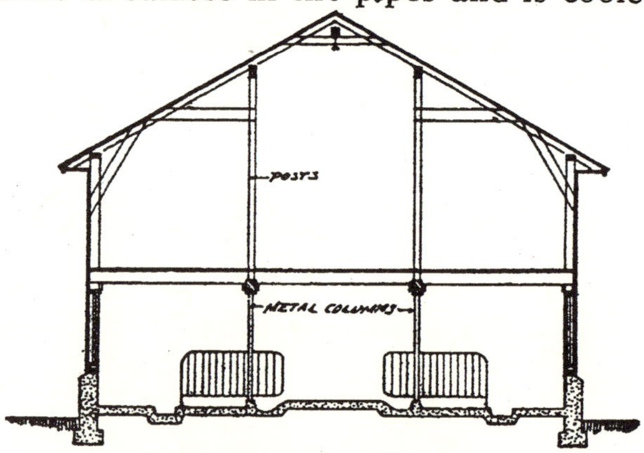

SECTION THROUGH COW STALLS

by a mixture of chipped ice and salt packed around the pipe coil in the ice chest.

It is one of the more scientific arrangements for the economical production of cold storage that works especially well where conditions are all favorable.

SOUTHERN COW BARN—A207

Cost of Blue Prints, $15.00

The cow barn herewith illustrated is designed for a warm climate with the view of obtaining good results as an investment. All rules of architectural proportion and design, as far as books are concerned, have been laid aside. The barn being located at a place where good returns are required and no architectural beauty called for, it

is a success because it meets all its requirements.

The walls of the basement or stock room

are built of cement blocks and the entire surface being of cement will keep the room cool in warm summer weather. The second floor, or hay room, is of wood construction and covered by a flat roof.

There are two rows of cow stalls; the cows facing each other, and between them is a hay rack built of one inch wrought iron pipes set six inches apart, the bottom of the pipes being imbedded in the concrete floor and the top of the pipes run into a wood rail. This feed rack, extending about three feet above the hay floor, makes it convenient for filling and at the same time gives it additional capacity so that it will hold several days' feed. At the foot of each side of this feed rack there is a feeding trough formed into the cement floor for feeding other foods and for watering. This is a very good arrangement, not only for its compactness, but as all the hay that is dropped by the cows falls into the trough and is afterwards picked up by the cow in

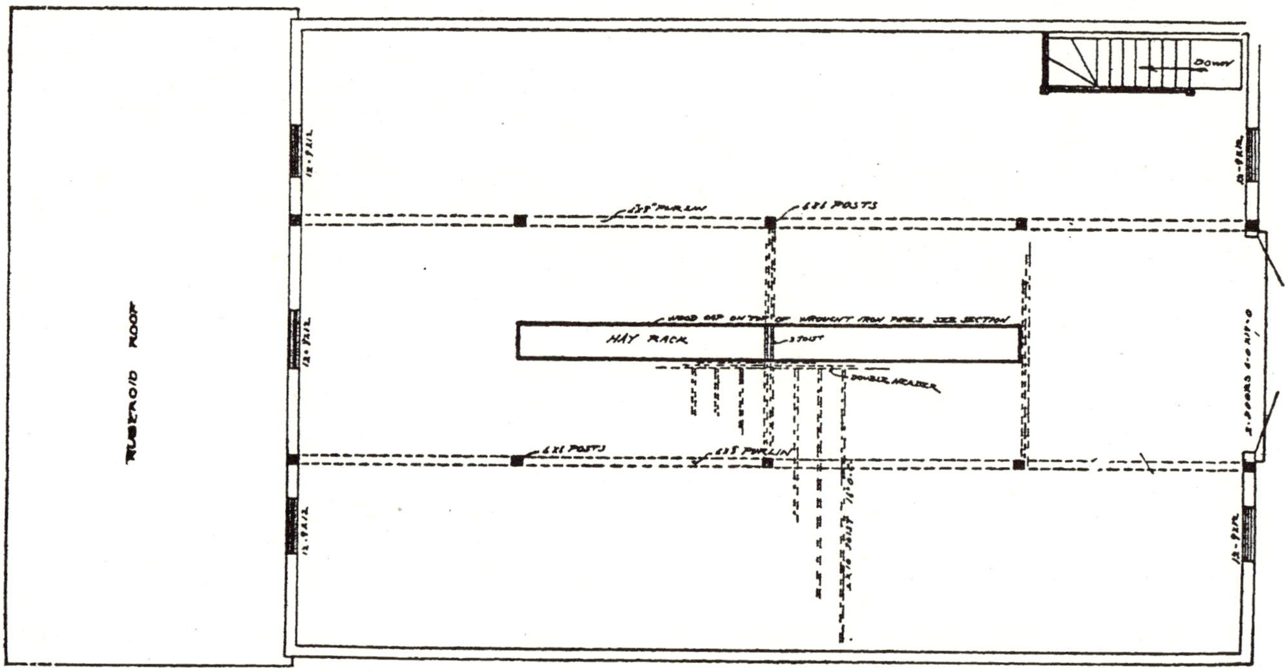

PLAN OF HAY FLOOR

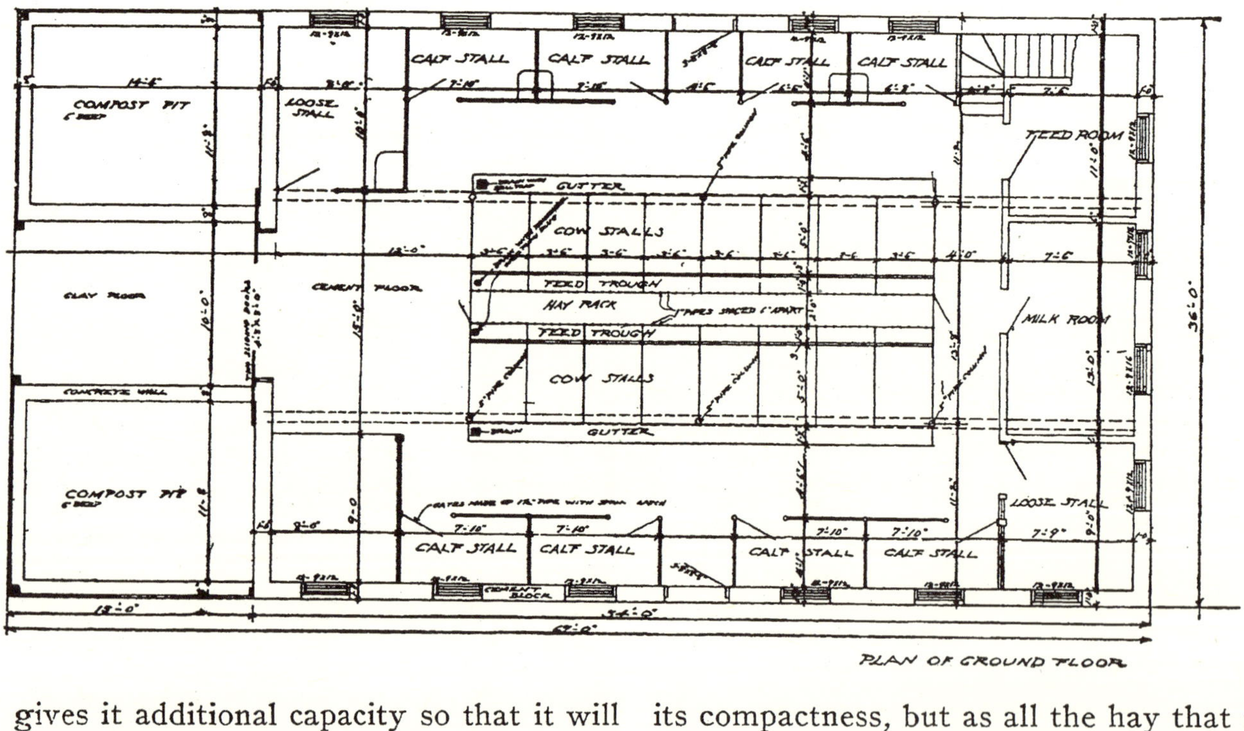

PLAN OF GROUND FLOOR

place of being tramped on and wasted.

The cows are separated by iron pipe railings which are imbedded in the cement floor and fastened to the hay rack.

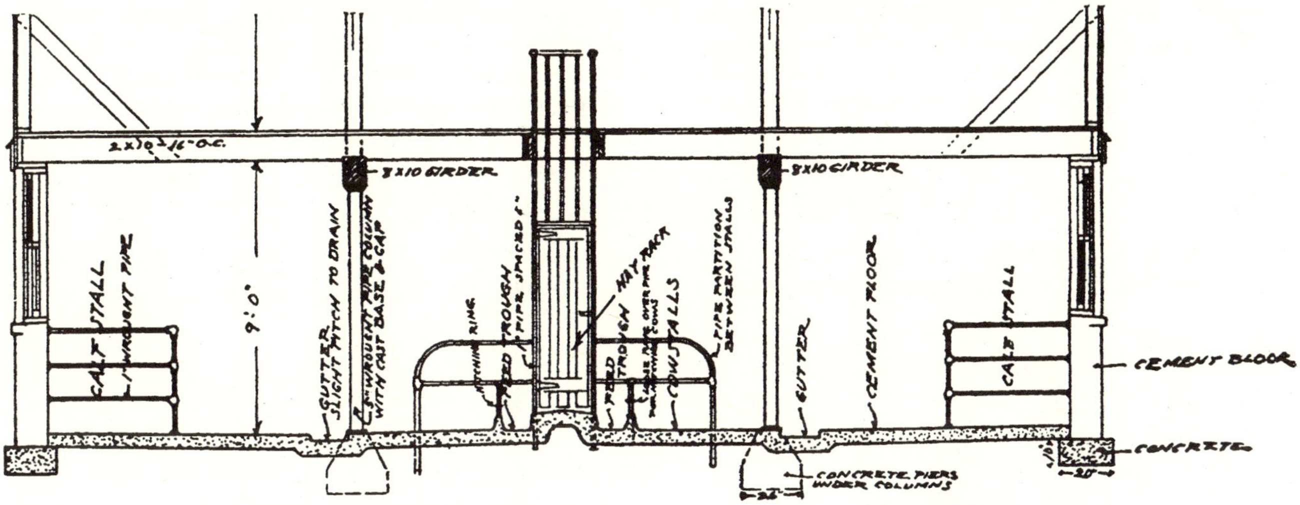

As will be noticed in the cut, the cows are fastened with chains which are fastened to a ring placed around a vertical pipe each side of the stall.

The stall floors have a gradual slope to a shallow gutter at the rear of the stalls and this gutter has a gradual slope to a drain tile to carry out all liquids to a cistern under the compost pits at the end of the barn. Calf stalls and loose stalls for dry stock and bull are arranged about the outside walls, and these are also constructed of wrought iron pipe, so that there is practically no wood work about the stalls or floors which can rot or get mouldy.

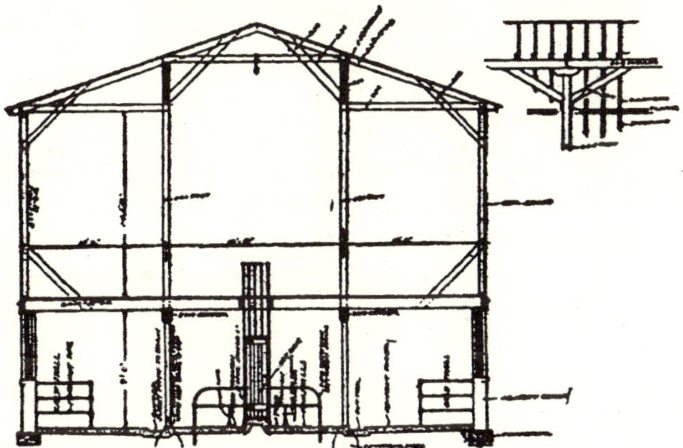

This makes an ideal barn for its purpose

and can be constructed at a reasonable price, and is practical for southern states where hay is the principal feed. For feeding silage this would not be so practical, as it would be too inconvenient to place the silage into the feed trough, unless the cows were first taken out of their stalls. Again it would not be well for a northern climate where the barn is constantly guarded against cold weather and a perfect system of ventilation and fresh air inlets are very necessary.

MODEL DAIRY—A180

Cost of Blue Prints, $15.00

We are here illustrating a dairy building which is very complete and answers heat, water, light and power for the various purposes required on a large dairy and

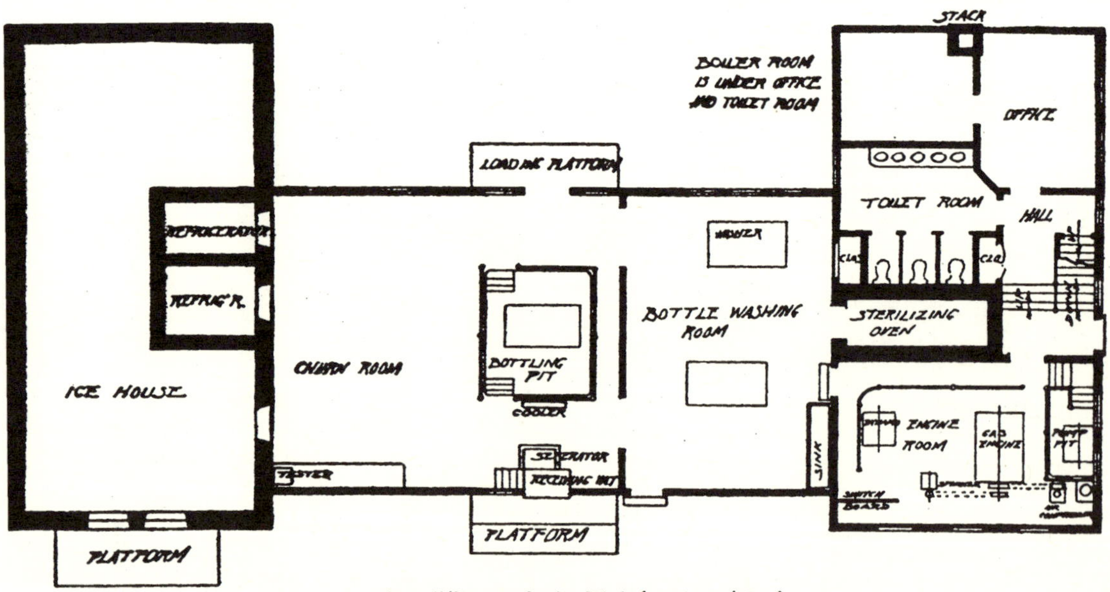

GROUND FLOOR PLAN OF DAIRY

all the requirements for a country dairy. It has waterworks, power and electric light plant of sufficient capacity to supply stock farm. The building consists of three parts; the left hand wing is the ice storage house and also contains two cold storage rooms for butter, cream, milk, etc.; the central part is the dairy containing the churn room, bottling room, washing room, etc., and the right wing is the power and pumping station.

This building is built on a concrete foundation, above which it is of the regular balloon frame construction. The walls are of two-inch by six-inch studding sheathed on the outside with matched sheathing, then papered and covered with drop siding. The space between the studding of the dairy and wash rooms from the floor to the window sills is filled with concrete and then cemented on the inside forming a cement wainscoting as well as strengthening the building. Above this cement work the side walls and ceiling are ceiled with beaded yellow pine ceiling. The roof is of moss green stained shingles and has large ventilators, which makes it hygienic and adds to the appearance.

The ice house is insulated with several thicknesses of hair felt, air spaces and matched sheathing and insulating, waterproof paper.

The power house has a basement which contains the boilers, which are sunk below the ground level in order to admit steam pipes to be run underground to the other farm buildings for heating purposes. The pumps and dynamo are run by an engine.

ANOTHER MODEL DAIRY—A176

Cost of Blue Prints, $10.00

A dairy building is located east of the cow barn and so arranged that the milk can be brought from the east door of the cow barn directly to the receiving vat in the dairy building. The milk cans are unloaded from the truck on to a platform, from which the milk is poured into the receiving vat from the outside of the building, thus avoiding the opening and closing of outside doors, which is very essential in order to maintain a uniform tempera ture in the building and to prevent the admittance of any impure air. From the receiving vat the milk flows by gravity through the various machines and apparatus without having to be handled by any hands until it is sealed in bottles, not only for economical, but more especially for sanitary reasons.

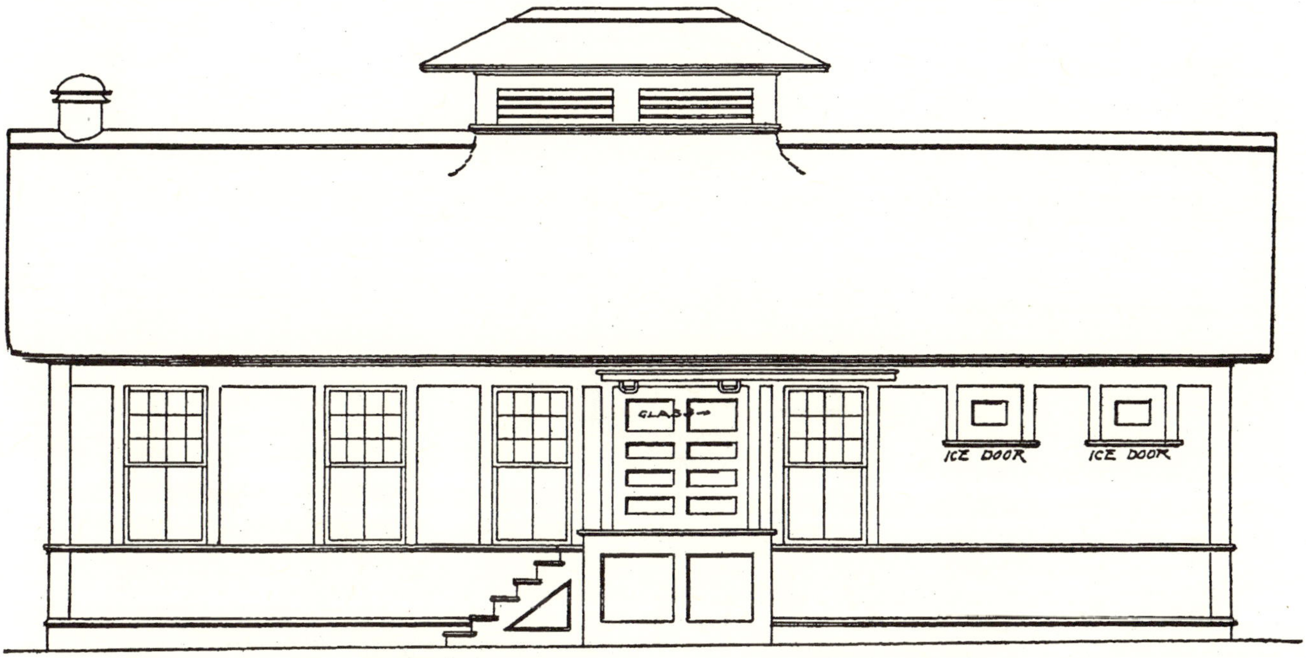

From the receiving vat the milk flows

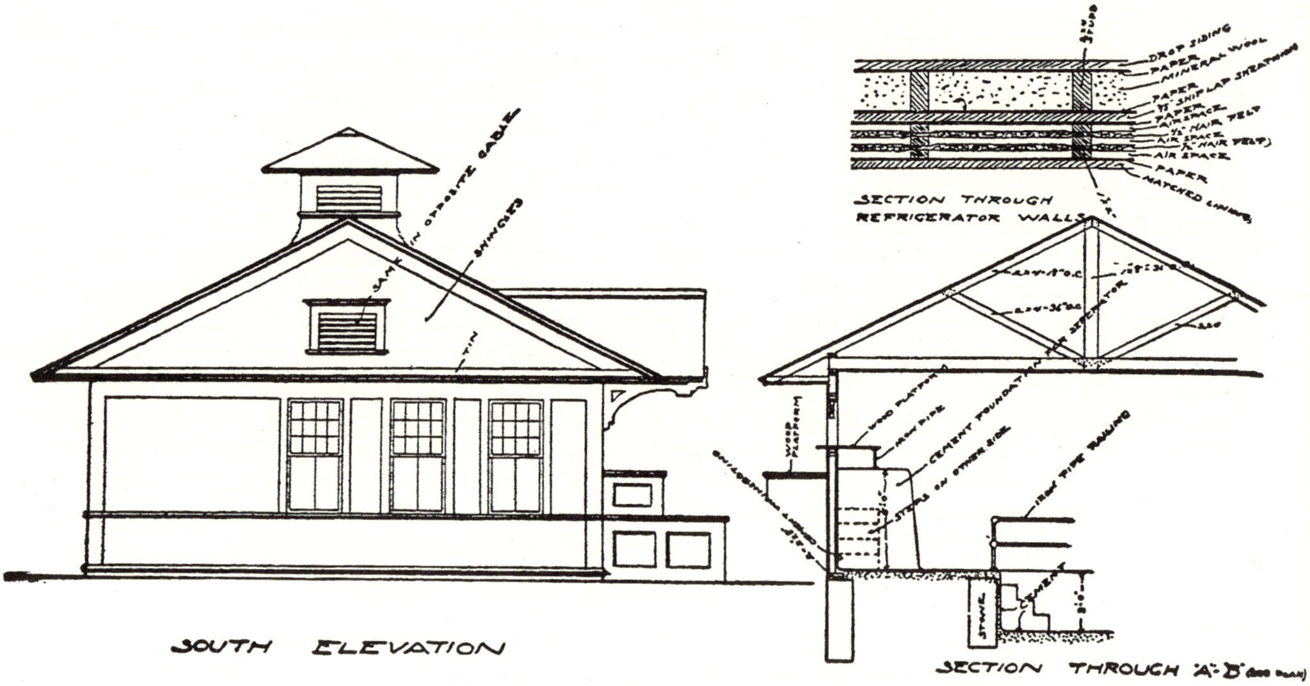

SUPPLEMENTARY DRAWING OF DAIRY BULDG.

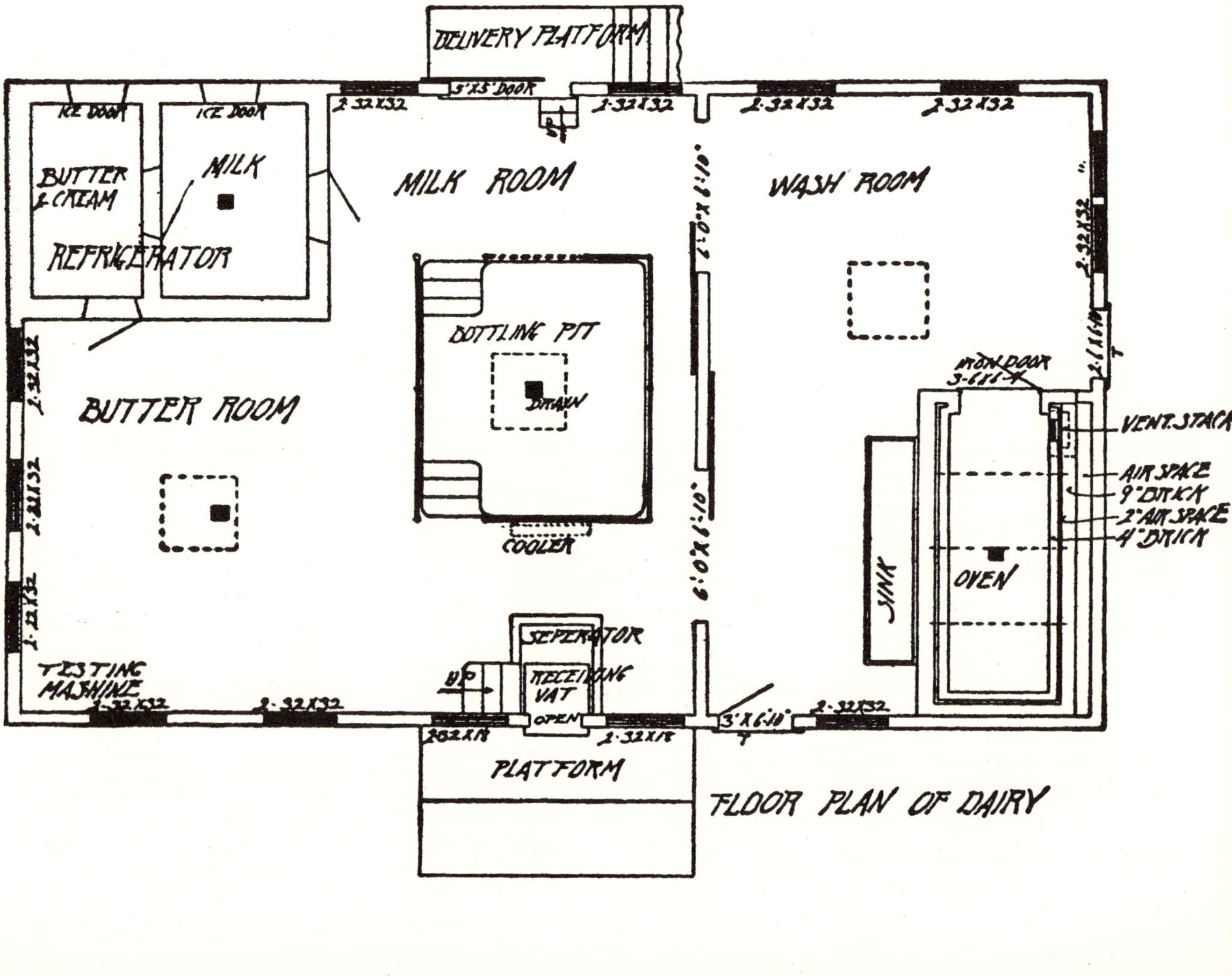

FLOOR PLAN OF DAIRY

into the separator and after the milk has been separated from the cream it is again mixed together and then flows through the cooler and into the bottling machine, which is located in a pit in the center of the milk room. The filled and sealed bottles are then placed into wooden delivery boxes for immediate delivery or else stored in the refrigerator ready for use.

In order to obtain a purely sanitary milk much depends on the care and cleanliness of the various receptacles, therefore too much emphasis cannot be placed on the washing and sterilizing. All the bottles are thoroughly washed by machines, which can do the work very thoroughly and rapidly by revolving brushes, etc., and after a thorough washing they are set into the sterilizing oven, which is equipped with steam coils and steam jets.

The butter room is located to the left of the milk room and is well equipped with the most up-to-date churns and also contains the testing machine and other apparatus. The refrigerator is divided into compartments, and is of the most approved construction.

The construction of this building is of the usual balloon type, having a stone foundation under walls of 2 by 4-inch studding, which are sheathed and sided on the outside. Between these is placed a double thickness of heavy building paper.

HOME DAIRY—A206

The very best butter is made on farms where the women thoroughly understand the business and have the proper facilities. When the milk from good healthy cows is run through a cream separator as soon as possible after milking and the cream cooled to the right temperature and kept in clean, pure air and churned when it is just old enough and not too old, you get the very nicest butter that skill and energy can make. Such butter, if shipped regularly to consumers in large cities, will bring a bigger price than the best creamery butter made. It is no more work to make butter right than to make it wrong; in fact there is less work because you have a proper system and that always helps.

The farm dairy may be very simple but is should be by itself. You cannot make gilt-edged butter in the kitchen. There are too many odors from cooking and sometimes from tobacco smoke. Cream is very touchy when it comes to odors. You can insult cream with a bad smell quicker than any other food product and when it is once contaminated no cleansing process can possibly eliminate the trouble.

The dairy may open off from the kitchen but you must keep the door closed. You will need some means of heating this room in winter time, but during the spring, fall and summer it will be warm enough without, and if it is on the north side of the building it will be cool enough most of the time without using ice, but ice is cheap enough to have and use when you need it.

This dairy is intended for from ten to thirty cows. If you have more cows you may need more room, but that will depend

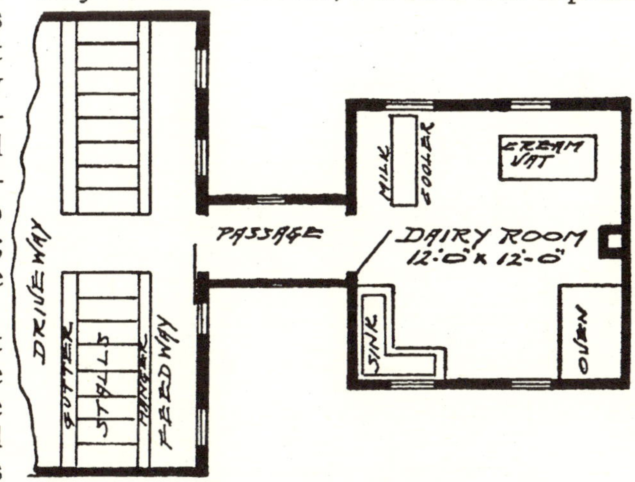

to some exten on how often you ship the butter.

The intention is to pack the butter in one pound prints or five pound crocks and to ship in neat little crates holding two or three crocks each or in boxes holding ten or twenty prints. If you have the proper storage and the butter is made to keep you can hold it in your store room until you get ready to ship it.

Department of
General Farm Barns

AN OHIO BARN—A146

Cost of Blue Prints, $10.00

A STYLE of barn that is very much used in Ohio is shown in plan (A146). A peculiarity of this style of barn is what is commonly termed a double threshing floor. In some of the larger ones the threshing machine is set first on one side and then on the other for convenience in getting the grain to the machine. The bridge from the bank to the ground floor must be stronger than common barn

bridges because it spans the space between the barn and bank and it leaves a runway for cattle along the bank side of the building. In this plan the cows have no stalls but are stabled in an enclosed shed with a feeding rack the whole length of the side so arranged that it may be filled from the mow above. Several removable racks for feeding grain may be placed anywhere in this shed and a water trough with an everlasting supply of good pure water will hardly freeze in here.

There are many points of convenience about a barn built after this plan, one of which is the facility of getting all around it. Gates, fences and retaining walls for the bank offer opportunities for stock pens in almost every corner without interfering with the barn proper. The entrance to the barn being overhead the whole ground space around the barn is left free to handle stock. Horses, cows, sheep and hogs may all have different quarters and be kept separate very much to the advantage of the stock and at a great saving in time. The dampness which is a bad feature of most bank barns is obviated in this plan because there is a circulation of air all around.

One of these barns was built on a hilly farm in southern Ohio on a site some distance from the house and about twenty feet higher, in fact the house was on one hill and the barn on another with a small ravine separating them. Two round wooden water tanks were placed near the top

SIDE ELEVATION OF BANK BARN

of the barn and these tanks were kept supplied by means of a hydraulic ram working from a running spring of pure clear water back among the hills.

To facilitate cleaning the tanks one at a time, they were connected at the bottom with a short pipe. In this pipe were two globe valves and between the valves was the outlet pipe to the house and to the stock watering troughs.

The pipe that brought the supply from the spring entered the tops of both tanks in a similar way. Two valves in the cross pipe permitted water to flow into either tank or both tanks as desired.

This arrangement of feed and outlet pipes made provision for emptying and cleaning either tank at any time without interfering with the water supply because the other tank could be continued in use. In practice it was found desirable to clean both tanks twice each year because if left longer they were inclined to become slimy.

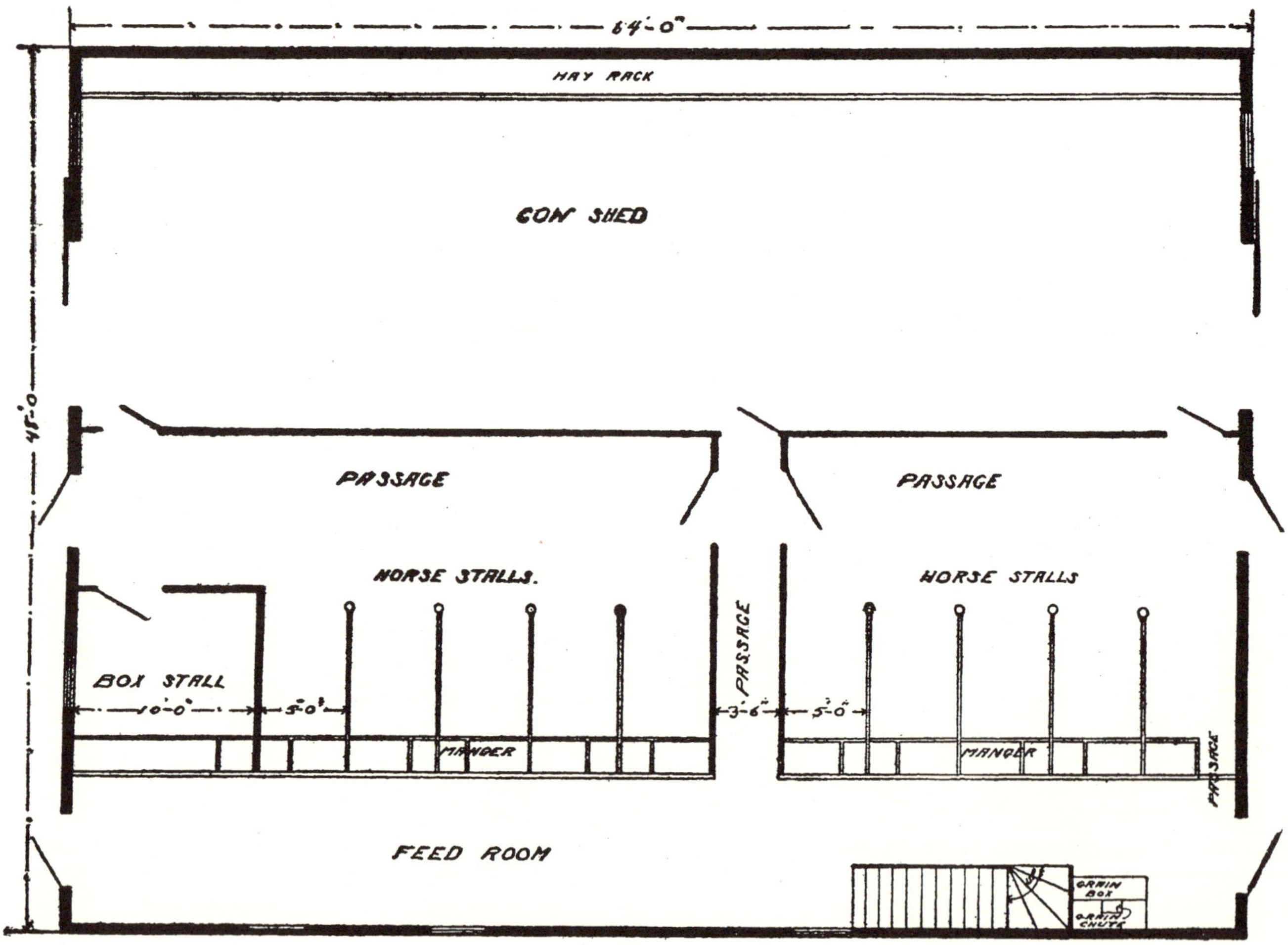

FIRST FLOOR PLAN

About seventy-five head of cattle and horses were kept on the farm besides other stock and their thrift was due in great measure to the unlimited supply of good water within easy reach at all times where they could drink out of cement troughs and cast iron buckets in convenient places about the stable and nearby pasture lots.

Besides supplying the stock an inch pipe was carried under the ground to the house, which was in this way supplied with hot and cold running water in the kitchen sink and bath room. There was also an outside hose tap for sprinkling the lawn and watering the flower beds. Another hose cock in the carriage house supplied a hose brush for washing buggies.

It might be noted that help stayed along on the farm year after year. One man grew up on the place from a chore-boy and only left to get married and work on a farm of his own. Farm hands are quick to appreciate modern improvements. Farmers who plan right can keep help and make money from their work.

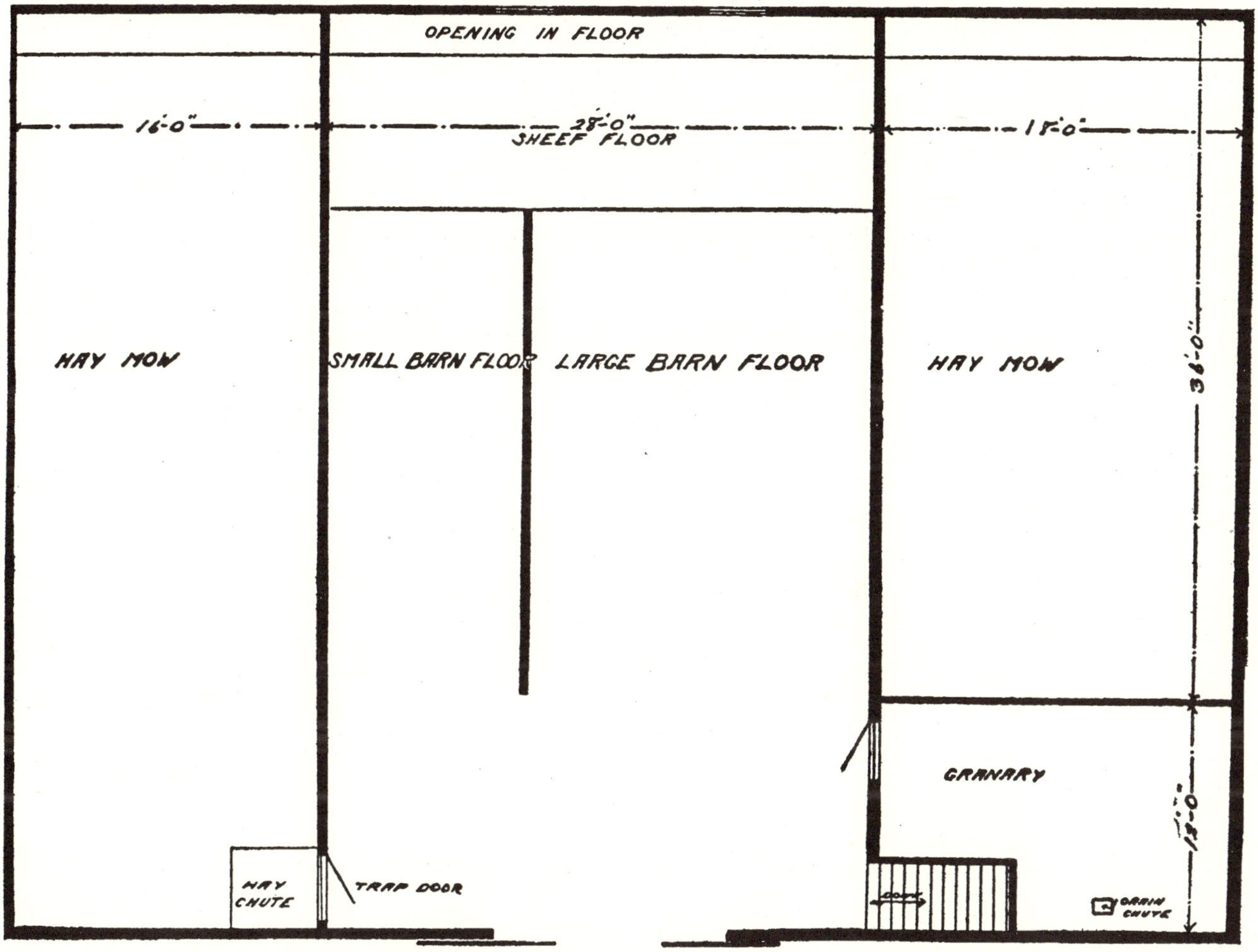

SECOND FLOOR PLAN

LARGE BANK BARN—A166

Cost of Blue Prints, $15.00

A bank barn is very desirable where a suitable location can be found but some bank barns are very inconvenient and others are damp and musty because the barn is not built right. It is not absolutely necessary to build a bank barn just because there is a hill on the farm. It is much better to pick out a plan which is suitable for the location than to blindly follow the lead of some other farmer. A barn that is all right on one farm may be all wrong on the next farm, so much depends on the use made of it, the kind of farming and the lay of the land.

This bank barn is 30 feet wide by 70 feet long with a basement full size. The walls of the basement are of stone and the upper structure is heavy frame work braced in such a way that a horse fork could be used in the peak with a track clear from obstruction extending from one gable to the other.

There is no objection to making this wall of cement or concrete if stone is scarce or if for any other reason a farmer prefers cement construction. This barn is placed sideways to the bank and has two bridges leading to what is commonly termed a double threshing floor on a level with the ground on the upper side. There are two doors on the opposite or south side of the barn but they are designed merely as openings for light and air as occasion requires and to run the carriers out when threshing. It is intended to build the straw stack in the yard on this lower side of the barn.

The basement is partitioned off into stables for six horses and twenty head of cattle as shown in the basement plan.

In building a barn like this it is necessary to use heavy timbers over the stable and to support them with good solid posts with good stone foundation or thoroughly well constructed cement footings solid enough to prevent settling. A good many such barns give considerable trouble in this respect but not necessarily so because it is easy to make them right in the first place.

In all stock barns, but especially where stock is kept in the basement, ventilation is of prime importance. This barn has two ventilators extending through the roof at the peak.

For convenience in feeding there are two chutes running down from the hay mow to the feed alleys on the stable floor. The double threshing floor leaves considerable room for storage of farm imple-

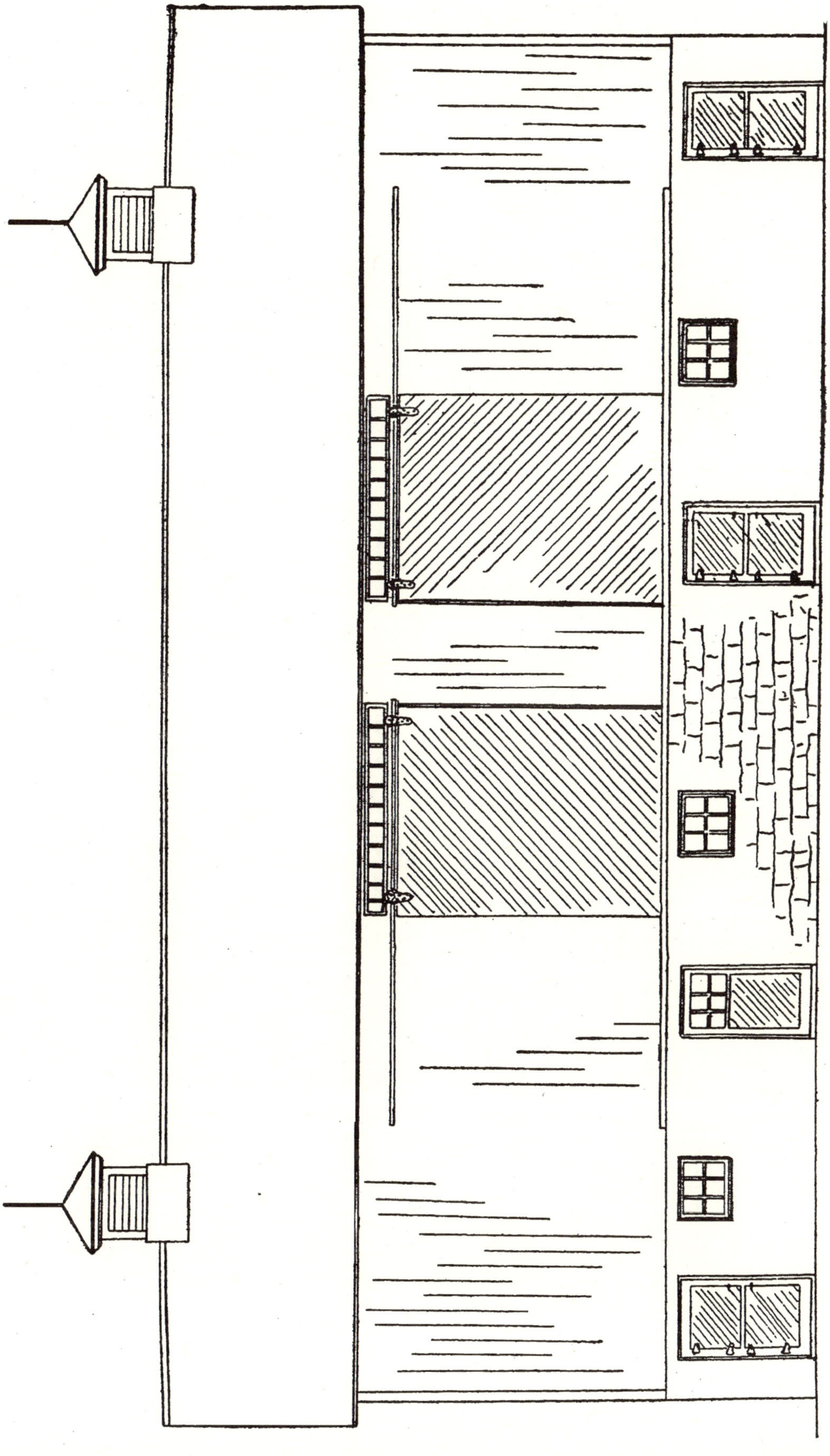

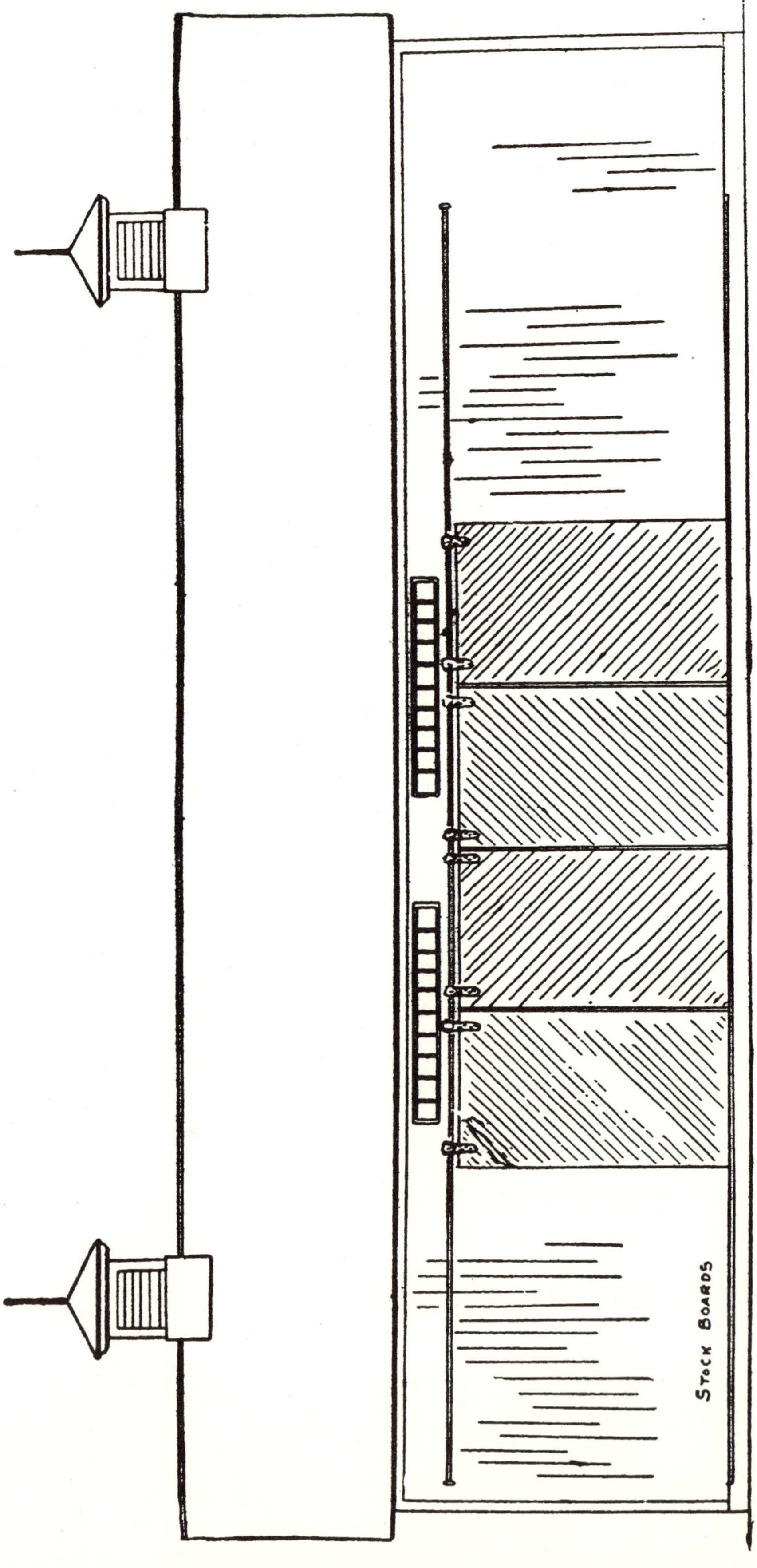
Stock Boards

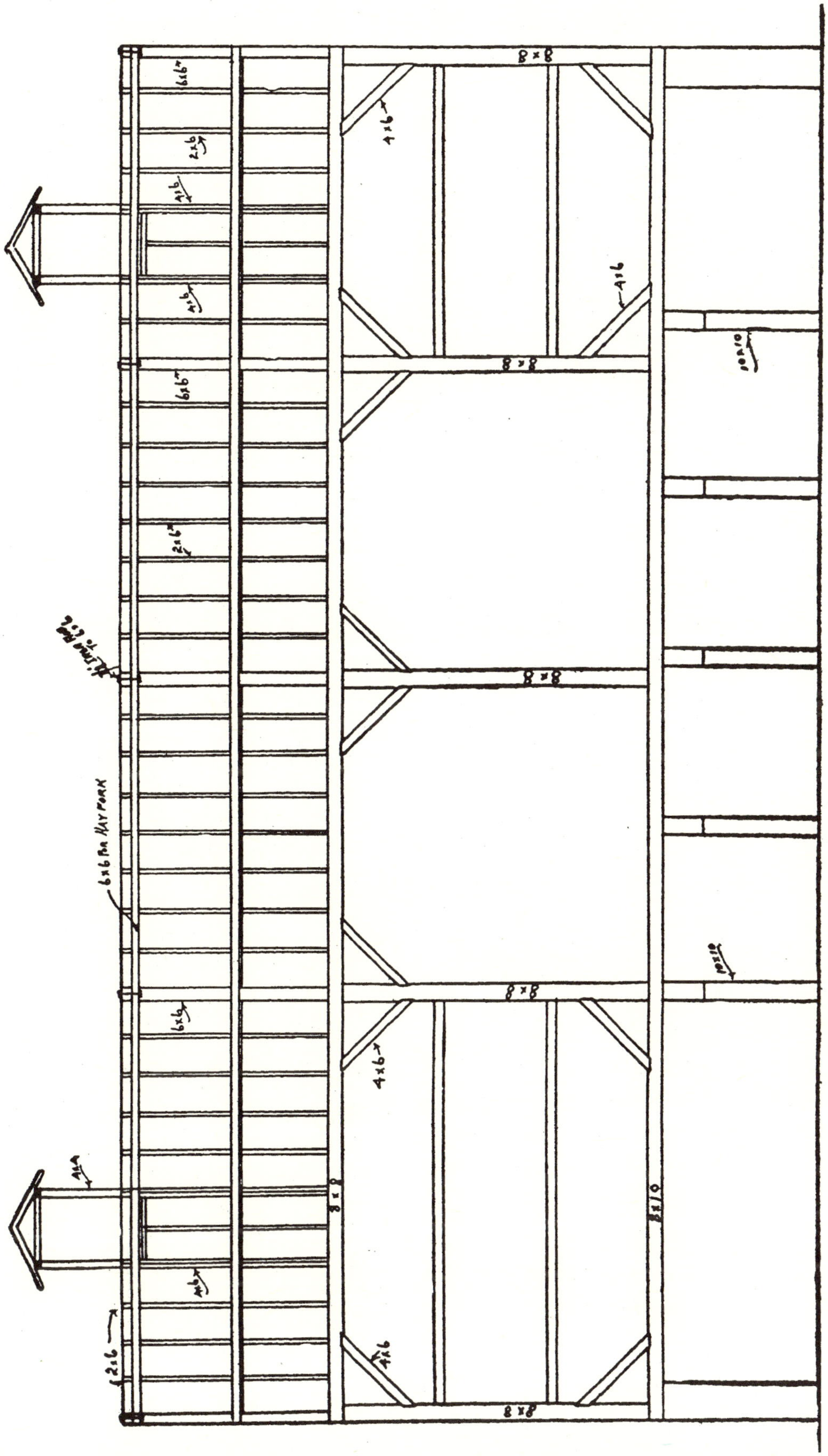
8 x 8
4 x 6
8 x 10

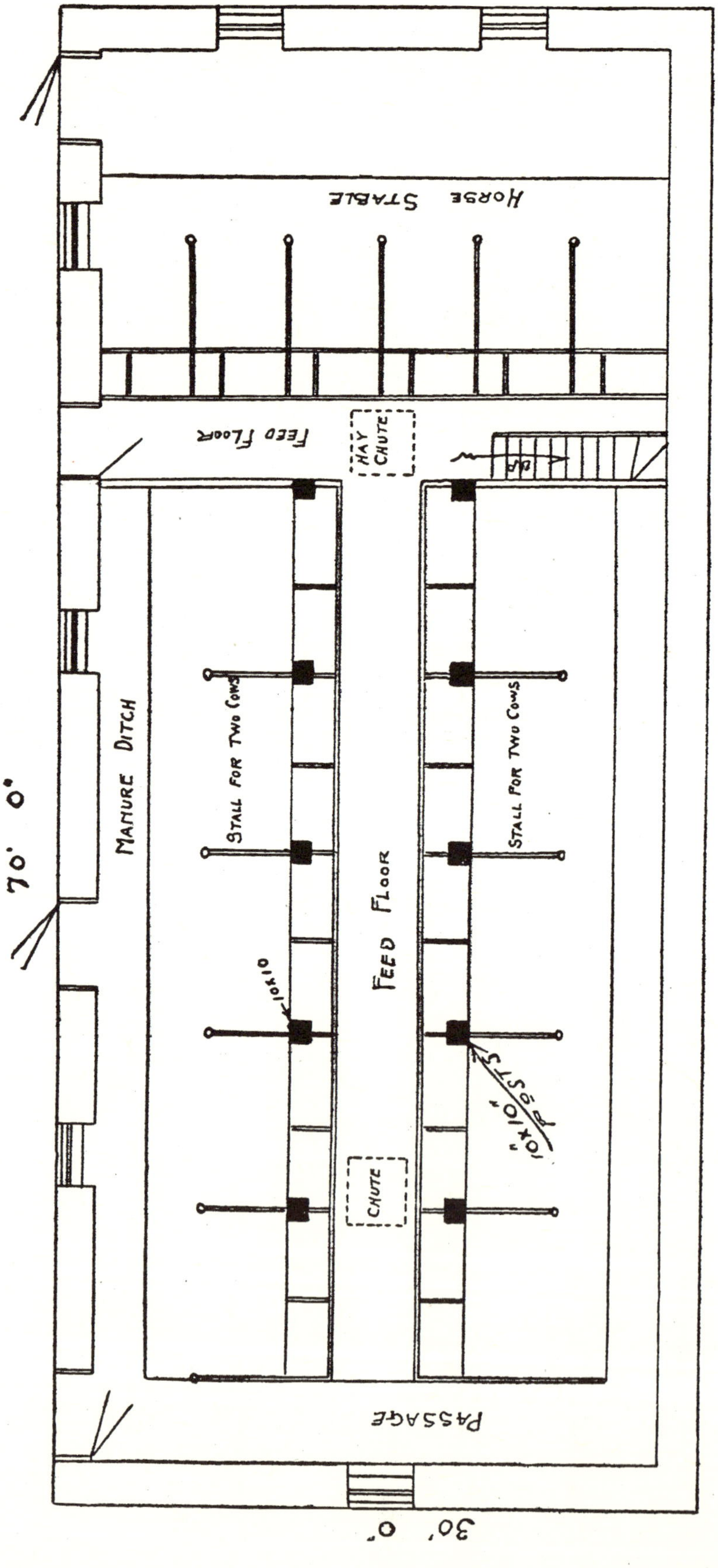
Horse Stable
Feed Floor
Hay Chute
Manure Ditch
Stall for Two Cows
Stall for Two Cows
Feed Floor
10x10
10"x10" Posts
Chute
Passage
70' 0"
30' 0"

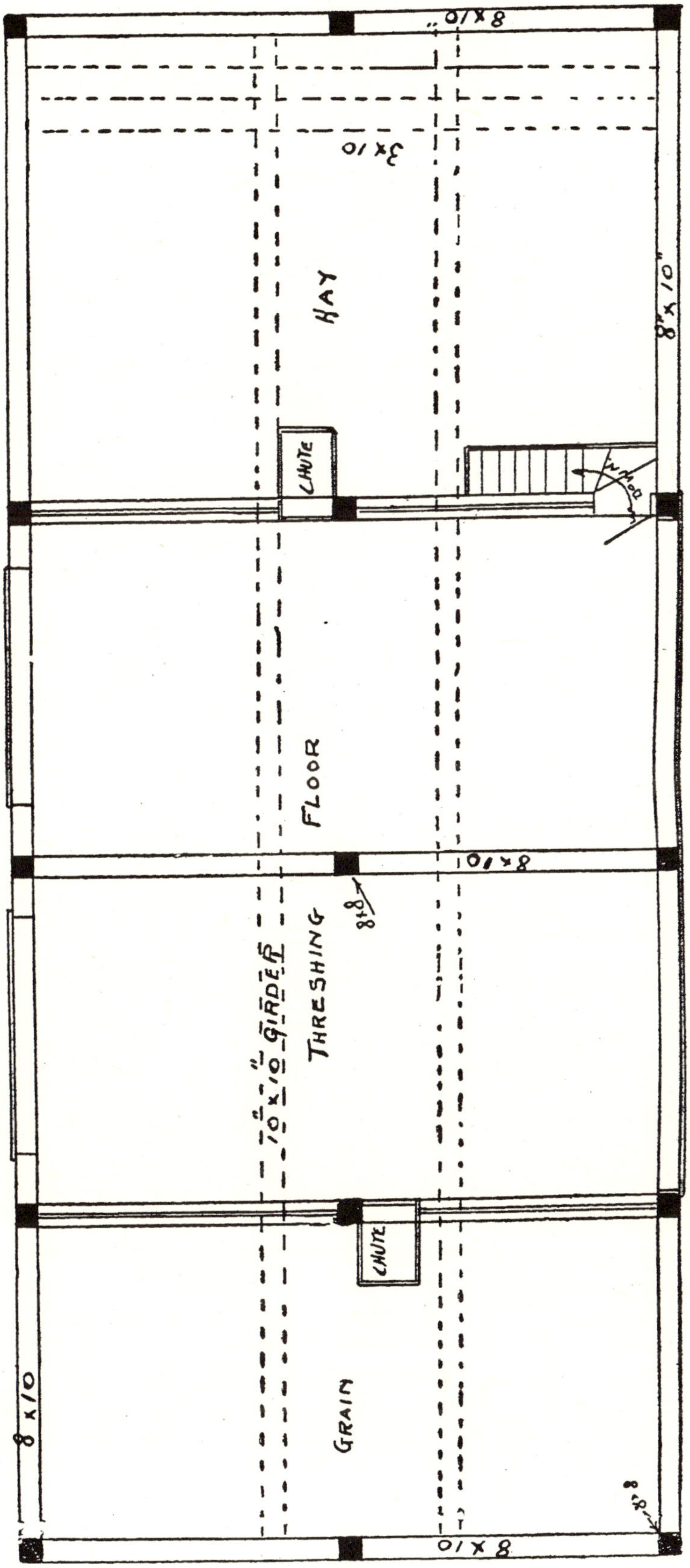
8x10
3x10
HAY
8"x10"
CHUTE
FLOOR
8x10
8x8
THRESHING
10"x10" GIRDER
CHUTE
8x10
GRAIN
8x8
8x10

ments which is very important on most farms. Where the land slants like this the barn yard usually is dry but probably a little tile draining helps every yard. We seldom see a barn yard dry enough in the fall and spring. It is well to consider all these side issues when selecting the site to build on.

BALLOON ROOFED BARN—A143

Cost of Blue Prints, $20.00

A good sized barn with a basement stable, a good threshing floor and a large storage for fodder is shown in plan (A143). The wall may be made of stone or cement according to circumstances. Eight feet head room is enough for the cow stable but usually nine feet is better for a horse stable. This barn should front the south and the root house should be, if possible, in a bank on the north side and the feed alley so arranged that a feed car may be run into the root house on a level.

It probably would be better to construct this case there is a good deal of outside wall clear of the bank and the windows may be made large.

Balloon roofs are becoming quite popular in barn construction, but some of the first ones were not made strong enough and heavy winds wrecked them. This roof however is braced by the gables from every direction which makes the structure a strong one.

The threshing floor is open in the center to the roof but it may be floored over at the ends if so desired. The intention is to

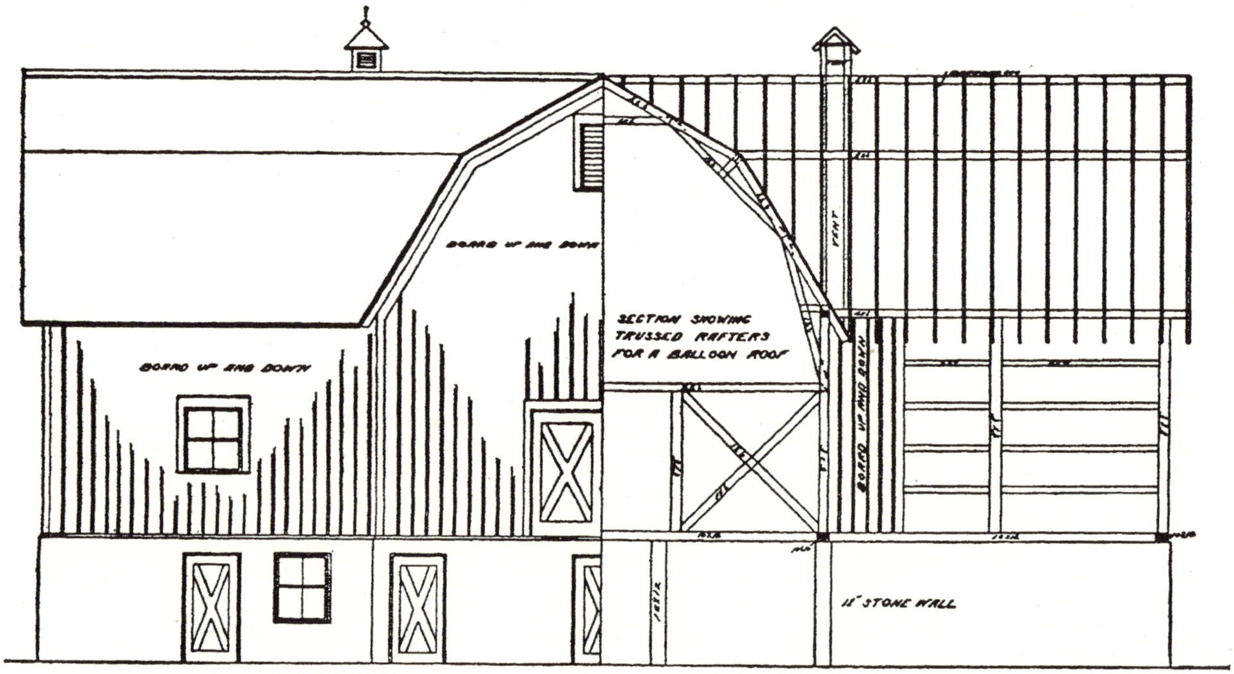

ELEVATION SECTION

a board partition between the horse stable and the cow stable, but the calf and bull pens would be better without a partition because the air will circulate better and there will be more light in the cow stable.

One objection to the basement stable is the difficulty of lighting it properly. A good deal depends on the exposure. In work the horse fork from this floor; to drive in with loads from the bank at the north and back out.

It is a good plan to leave sufficient opening to run the straw carrier or stacker up to the mows above. On most farms it would be desirable to have a stack in the yard but it is just as well to put some of the straw back in the barn. A balloon roof works splendidly for this purpose. The stacker may be turned to blow the straw to the furtherest end of any gable.

It is a good plan to pay careful attention to the ventilation of any stable. The air in a basement stable is seldom as good as it should be. There are two air shafts in this plan with openings near the floor.

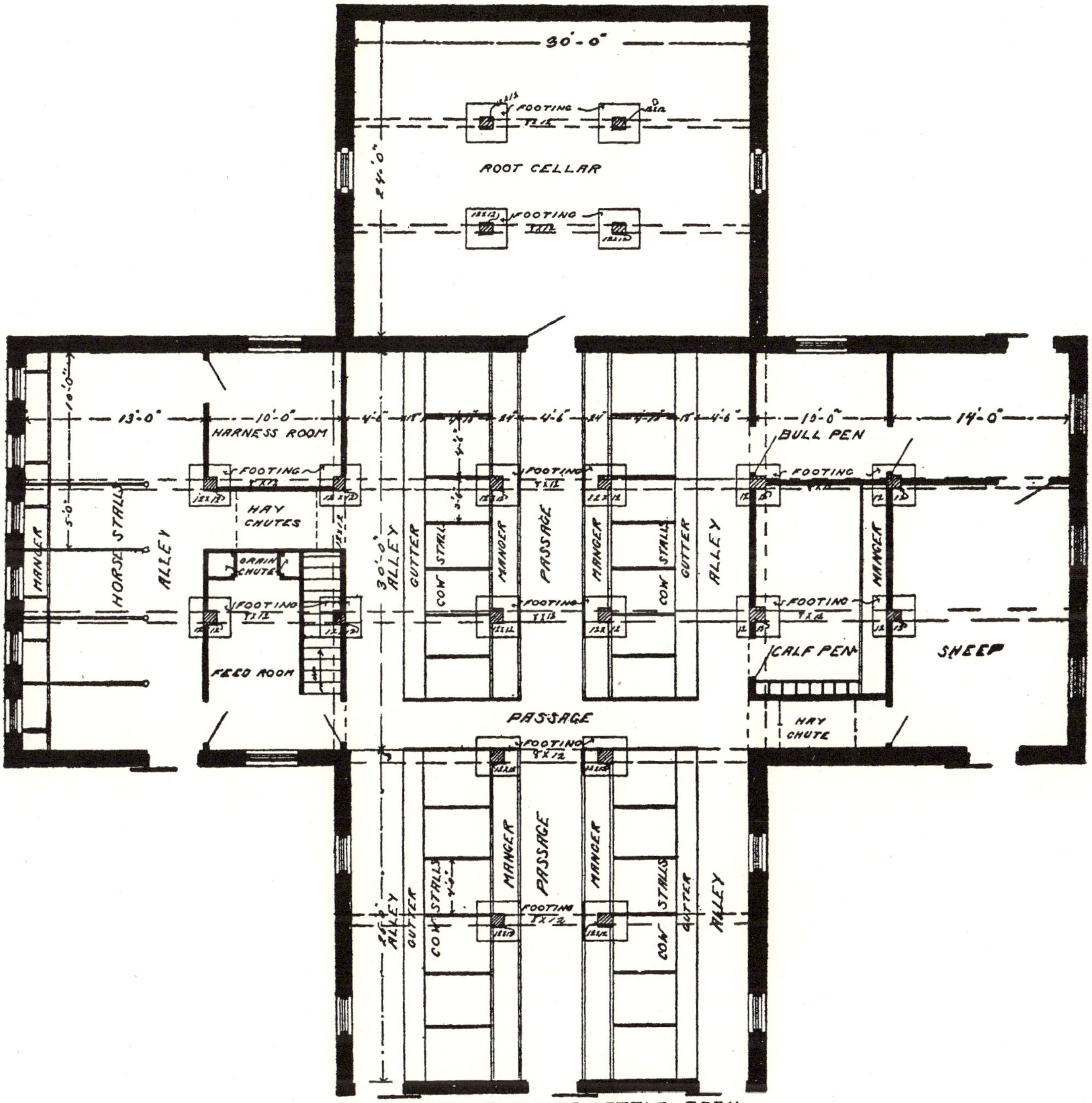

FLOOR PLAN OF CATTLE BARN

It will be noticed that two hay chutes are provided to carry the hay down to the feed alleys. Hay chutes are a great convenience but they are draughty things unless doors are provided. In putting in the upper floor timbers and joists it is a good plan to make them continuous by building them up with two inch plank so as to tie the building together in both directions.

Remember in building this barn you have no upper ties and you must support the roof from the frame below, but this is easily done because of the shape of the building.

Some farmers may need a larger granary than the one shown in the plan. In that case it may be extended to cover the whole floor in the granary wing, which should make the granary about twenty-two by thirty feet and the hay shoot would pass down through it just the same.

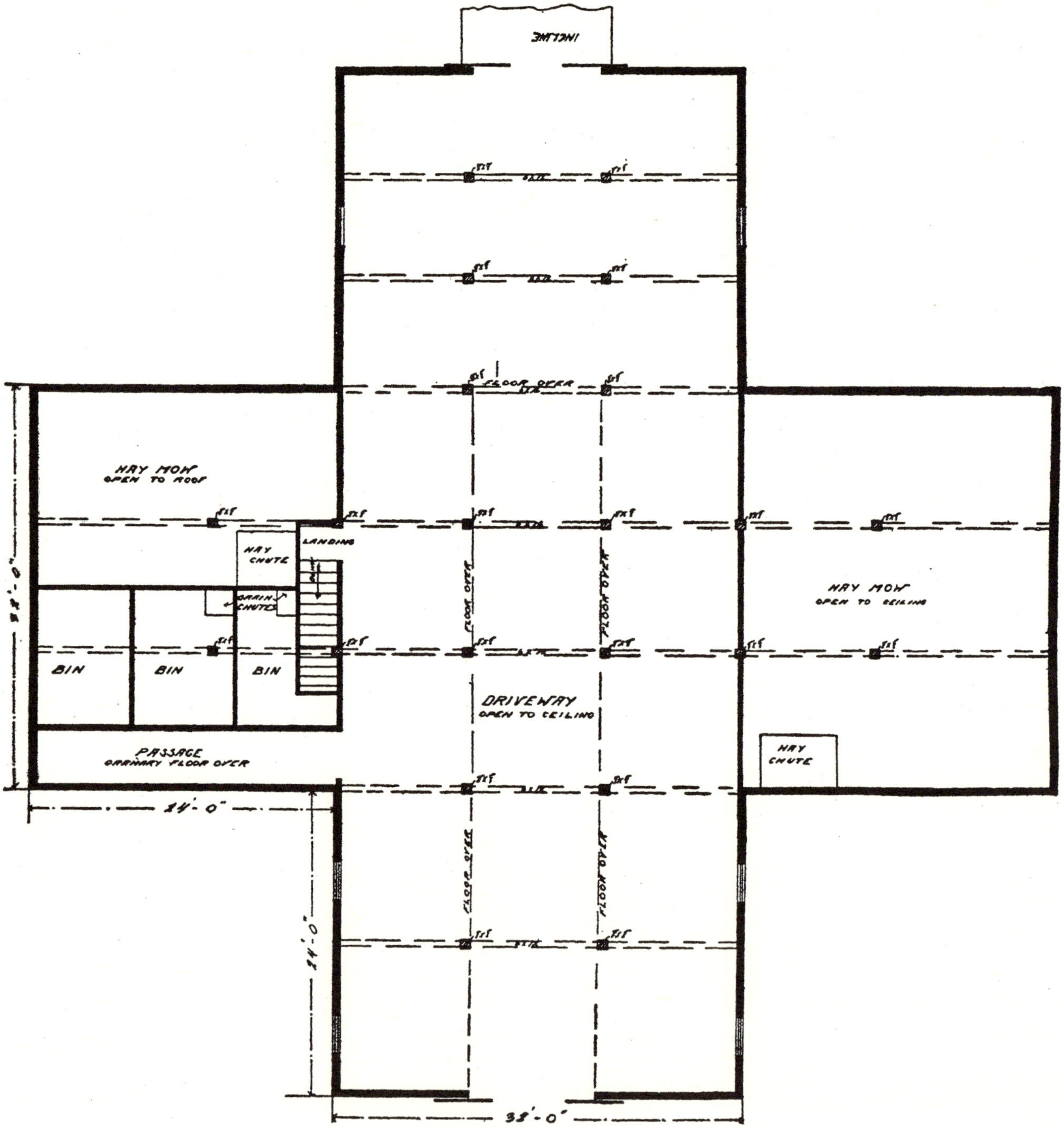

SECOND FLOOR PLAN

CANADIAN BARN—A183

Cost of Blue Prints, $250.00

This large stock and dairy barn was designed for a large Canadian farm and has many good features worth noting, both from the builder's and the dairyman's point of view.

The shape of the building was developed with the view of giving the best shelter to the stock. From the points of the compass, as shown on the floor plan, it will be seen that the wings of the cow barn and the young stock barn are so situated as to keep the north wind off the stock when it is let out for exercise during the winter months, and at the same time giving them all the sunshine. The building is also arranged to be convenient from the paddocks, pastures, etc., allowing the stock to approach their respective stalls without having to be driven across unnecessary driveways or through a series of gates.

The building is built of wood, on a foundation of concrete, which is put in place by excavating the trenches the exact width and depth of the wall and then the concrete is dumped and tamped into the trench, thus avoiding the work and expense of planking for concrete forms below grade. Above grade the concrete is tamped between planks well fastened in place in the usual manner. The concrete wall extends up to the floor level where the wood construction begins. The space between the studding from the floor up to the window sill level is also filled with concrete after the walls have been sided with drop siding over a layer of thick tar paper. After the concrete between the studding has become hard metal lath are put in place on the interior face of studding and over the concrete, which is then plastered with cement mortar, making a cement wainscoting around the walls, which makes a perfectly sanitary barn. The concrete filled walls help greatly to keep the barn warm in winter and cool in summer, as well as to stiffen the structure against heavy winds.

The granary is located at the center of the north side and contains eight large hopper bottom bins for the storage of grain and feed. The bottom of each bin is connected with a spout leading to an elevator boot in the basement, which elevates the grain to a revolving head so that the grain can readily be transferred from one bin to another or onto a truck or wagons. Some of the bins also have spouts wagon-bed height above the floor for feeding purposes. The main driveway of the barn goes through this granary and contains a combination dumping scales with a hopper under the floor spouted to the elevator boot for loading grain into the bins.

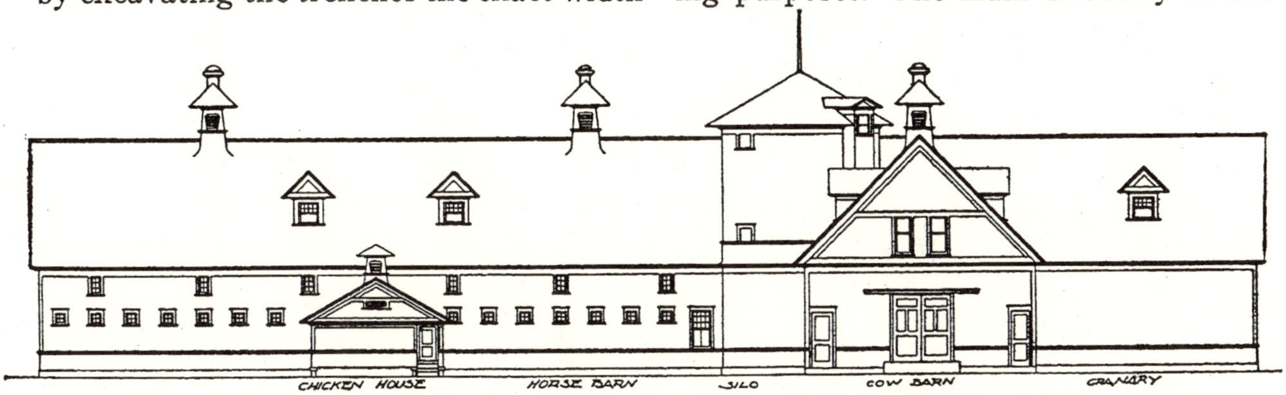

This granary being located near the center of the barn is very convenient for feeding the stock and adds to the exterior appearance of the building. The basement of the granary is used for the storage of roots for the stock and can be equipped with a kettle for boiling and mixing foods, etc.

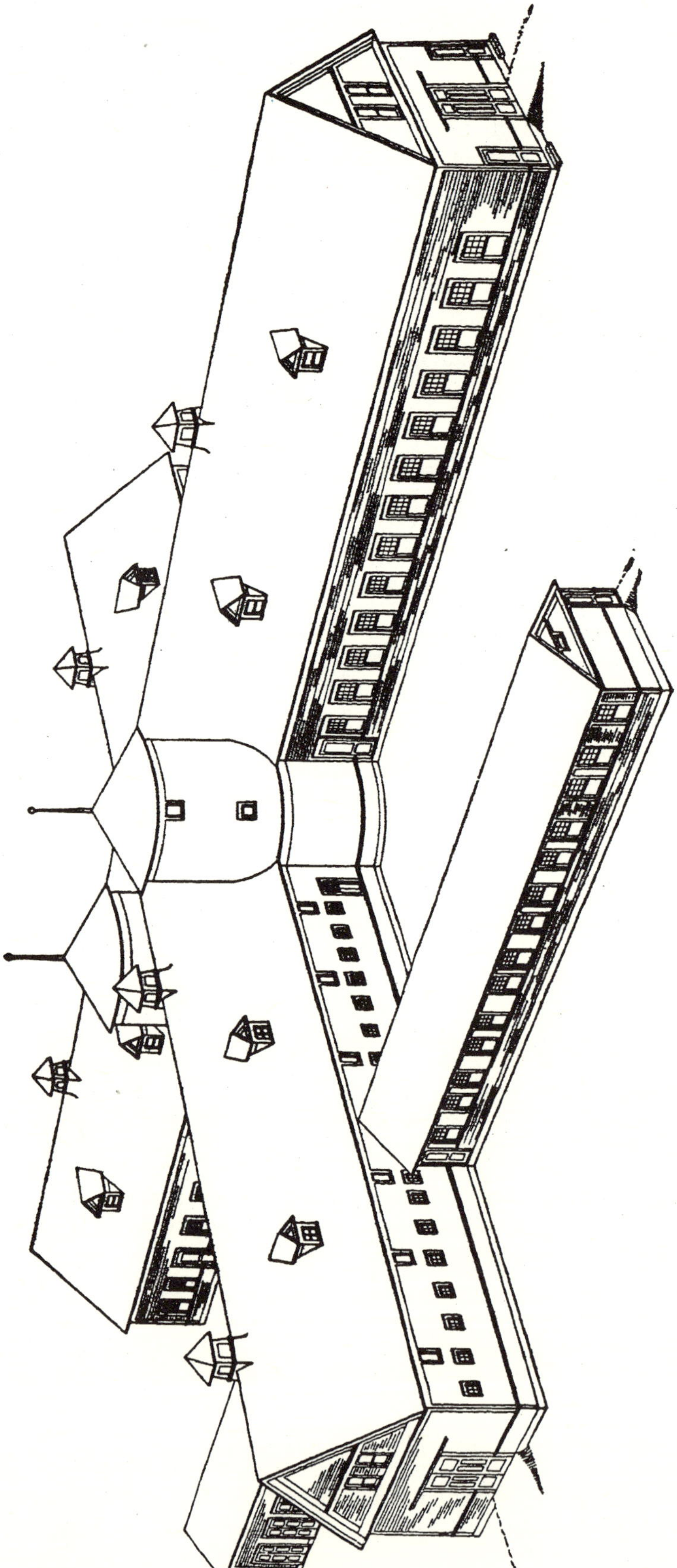

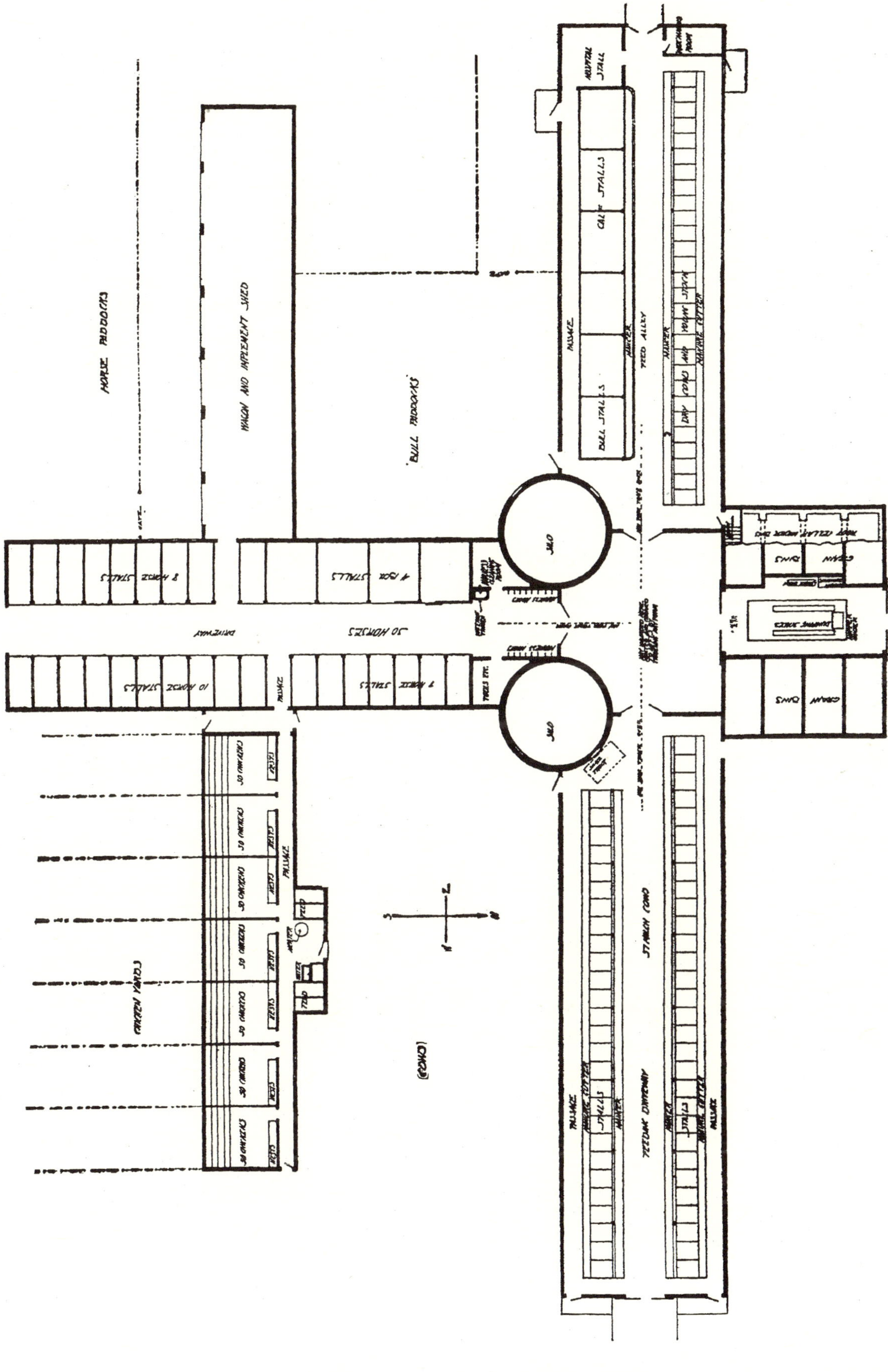
HORSE PADDOCKS
WAGON AND IMPLEMENT SHED
BULL PADDOCKS
CALF STALLS
BULL STALLS
PASSAGE
MANGER
FEED ALLEY
MANURE GUTTER
HOSPITAL STALL
SILO
SILO
8 HORSE STALLS
4 BOX STALLS
20 HORSES
DRIVEWAY
10 HORSE STALLS
7 HORSE STALLS
PASSAGE
GRAIN BINS
GRAIN BINS
FATTEN YARDS
STANCHION COWS
FEEDING DRIVEWAY
STALLS
(ROAD)

The cow barn contains 57 cow stalls and arranged with a feed alley running through the entire length with the mangers on either side, so the cattle can be conveniently fed from a truck or a trolley track system suspended from the ceiling. The cows stand facing each other and the mangers are continuous, constructed out of concrete which forms part of the cement floor. The stall floors are of concrete covered with plank, which can be taken up and cleaned or renewed when desired. The manure gutters have sufficient fall to drain all liquids to one outlet in the center which is connected with a catch-basin, and also contains gate valves so arranged that while scrubbing the water can be switched into a sewer. The passages back of the cows are of good width for milking and bedding the stock and trucking ont manure to platforms built at the end of each passage outside of the building. The ventilation is well taken care of by ducts in the walls which carry the air to the ventilators on the roof.

The young stock barn is located to the west of the cow barn and contains six box stalls for bulls and calves. These stalls are constructed from heavy wrought iron gas pipe, having three-inch pipes for corner posts and for top or header rail, and 1½-inch pipe spaced 6 inches apart for the stall partitions; these pipes are set upright with the bottom ends well bedded in the concrete floor and the upper ends screwed into 3-inch header. The gates are also of pipe construction and have self-closing locks and hinges.

There are 28 single stalls with swinging stanchions for calves, one-year-olds, and dry stock similar in arrangement to the stalls of the milk cows only not so wide, as no milking room is necessary.

The wing also contains a hospital stall which is isolated from all others by solid walls and has all side walls, floor and ceiling finished with cement which is impervious to moisture and can be readily disinfected. Opposite the hospital stall is a watchman's room for a man who can attend any sick stock during the night.

The silos are centrally located for convenience in feeding and filling, as the silage cutter can be located in the central feeding room and thus be operated in all kinds of weather during the ensilage season. The silos are constructed of studding spaced 12 inches on centers, sheathed on the inside with two thicknesses of 1½-inch by 6 inch sheathing bent around horizontal and then veneered on the inside with hard, vitreous paving brick laid in cement mortar, each brick being tightly pressed against the sheathing so that the silage pressure cannot force it out of place. The exterior of each silo is finished to match the balance of the building. The silos have a concrete foundation which is flush on the inside with the face of brick lining, and being excavated down to the footing increases its capacity by about 50 tons. The floors are of concrete, dished to the center, and connected with a deep seal trap and drain.

South of the silos is the horse barn, which contains 17 single stalls on one side and 9 single and 4 box stalls on the other side, giving it a capacity of thirty horses. Each stall has an outside window for light and ventilation. These windows are about seven feet from the floor to avoid draft on the animals and protected by a wire mesh guard. The stall partitions are of wood to a height of 5 feet 6 inches, giving a good circulation of air and light. The stall floors are of double thickness of 1¾ inch by 6 inch flooring with several thicknesses of roofing felt laid in hot tar between. All stall floors are slightly sloped down towards the driveway and have cast iron gutters with perforated cast iron covers and connected with catch-basin and sewer.

East of the horse barn is the chicken house, having a capacity of 350 fowls, divided into seven compartments of 50 each, so arranged that the chickens get the south sun and protected from the cold north winds.

East of the horse barn is the shed for wagons and farming implements with a door into the horse stable, so the team can be taken directly from the stable into the shed and hitched up without having to go through a barn yard.

There are many other conveniences about this building, but we must refrain in this article for lack of space. Suffice it, therefore, to conclude in stating that the building is so constructed that any department of the same can at any future time be extended or added to.

AN OCTAGON BARN—A150

Cost of Blue Prints, $15.00

This is a cement silo with a barn built around it. The arrangement is a good one for feeding young cattle to make them grow, rather than to fatten steers for the market. The silo is sixteen feet in diameter and thirty-two feet high with a twelve inch cement wall and a pit that reaches three feet below the surface of the ground. Three feet is deep enough to give a good solid foundation and it is deep enough when you come to pitch the last silage out of the bottom.

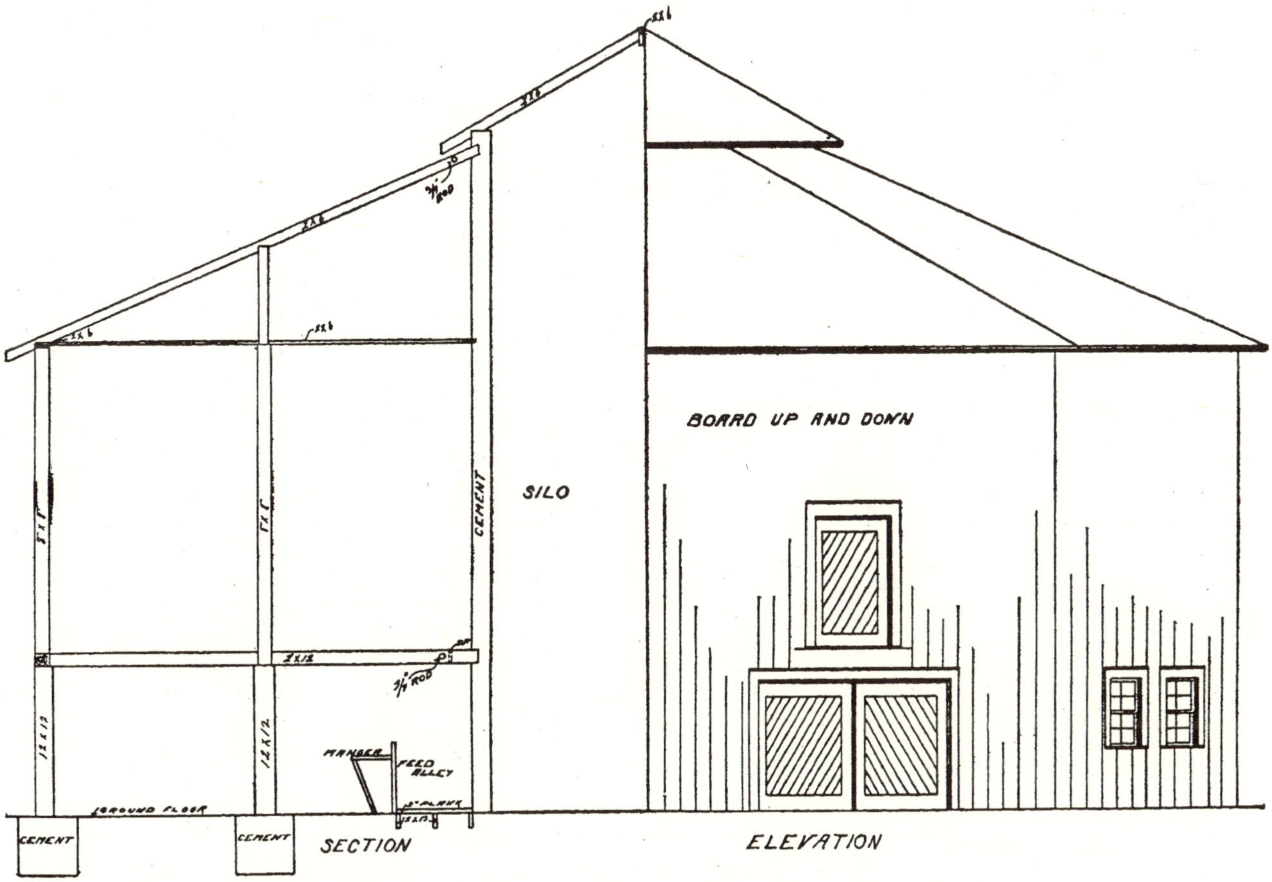

The frame-work of the barn is very light. The silo is used to support the middle and the barn really is braced from every direction. Every side is both a brace and a tie for the next side. To prevent any possible pulling away from the silo, rods connect all the floor joists and all the rafters. This makes a circle of three quarter inch iron at the floor and again at the roof, but if the different sides of the building are well tied together there will be no getting away even if the iron rods are not used.

The octagon construction has been worked out in this plan in preference to a round barn because the construction is cheaper. The sills and other timbers are straight. The joists usually are cut square, at least there are not very many bevels and when a joist is beveled it is only on one end and the other end is cut square. It is the same with the rafters.

There is considerable room for straw and hay around the silo and it is easy to make places next to the silo for putting both hay and straw down into the feed alley.

The mangers being next to the feed alley makes feeding as easy and convenient as it is possible to have it. Perhaps no other barn construction can offer such advantages at feeding time. The mangers hold hay, corn stalks or other roughage and the bottoms are tight for feeding corn or ensilage. The feed room in front of the silo doors is boarded to the ceiling so that

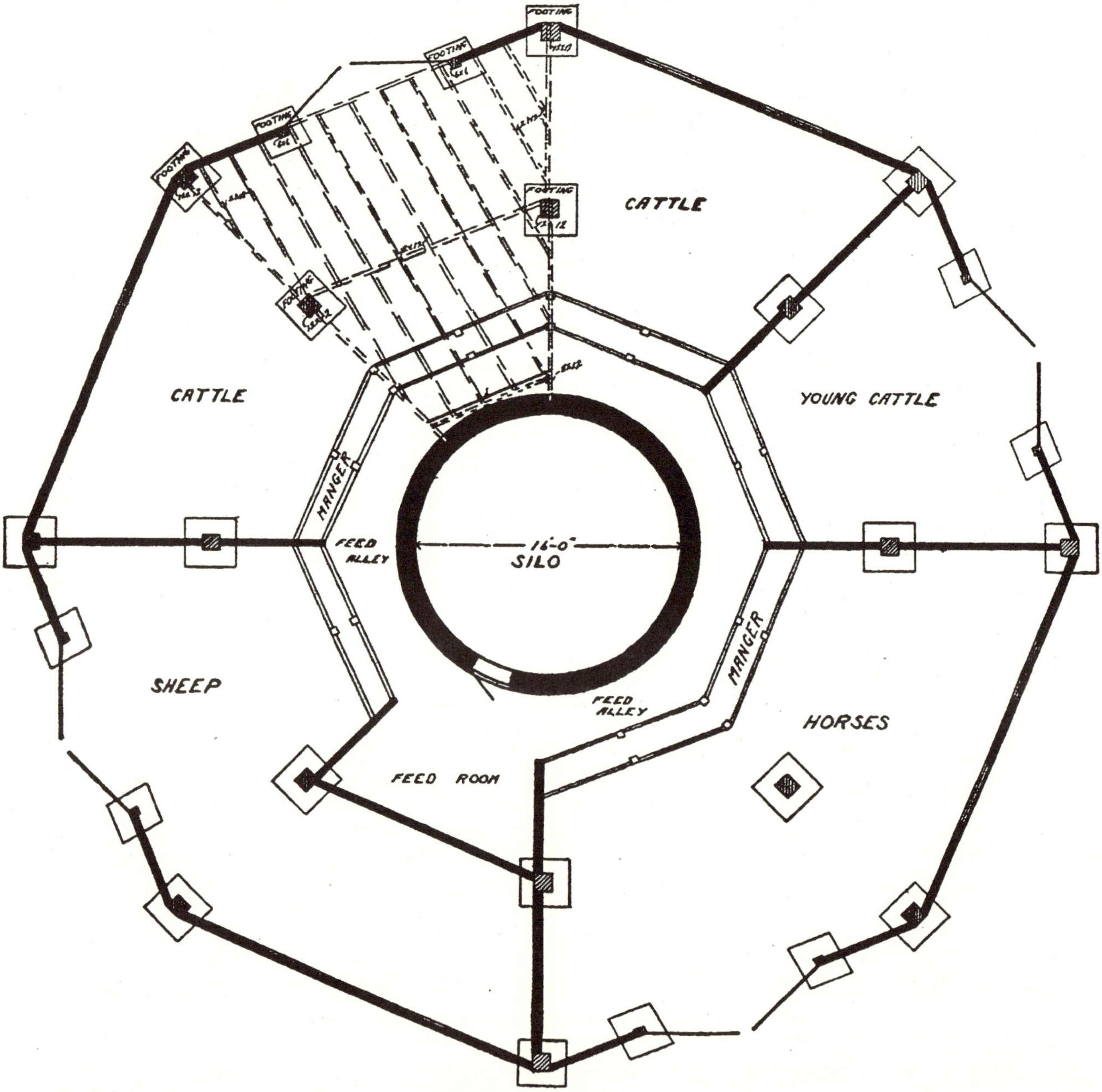

FLOOR PLAN

ensilage enough for a full feed may be piled up out of the way of the ensilage cart. A packing box with large castors may be used for a silage cart or it may be a well built cart with heavy iron wheels and with hinged sides to drop over to the manger.

There are four entrances for convenience in getting out the manure and most of them will be used at times for letting stock in or out, especially if the barn is divided up in compartments for the different kinds of stock. Each post has a good cement footing as shown in the plan and the elevation shows the way the timbers run.

There is no floor in the bottom except the ground as it is intended to let the straw and manure accumulate, but there is a good feed room floor as this is where the work is done three or four times a day. A silo surrounded like this must be filled with a carrier. A blast stack will not work well on an incline and it is not convenient to place the cutter close to the silo, but a good carrier works all right.

EIGHTY ACRE FARM BARN—A211

Cost of Blue Prints, $10.00

A general purpose barn thirty-two by sixty feet with storage room overhead may be built on this plan. The idea is to provide a barn that is suitable for a farm, say of eighty acres, where six or eight horses are kept for work and for breeding purposes, and where there is a variety of cattle, some milch cows and some growing calves and young stock.

The cow part is partitioned off from the other part of the barn with stanchions facing out. There is a manger between the stanchions and the outside of the barn with a rack to hold hay and there is a long narrow opening above this rack into the mow the whole length of the rack so that hay may be put down from overhead and distributed as it is thrown down, which saves once or twice handling when compared with feeding arrangements in some other stables.

A manure carrier runs on a track behind

the cows and the same track is extended to run behind the horses with a switch to throw it from one track to the other as needed. Carrying the manure all out at one door leaves the outside of the barn clean on three sides which adds very much to the appearance.

The main entrance is into an alley twelve feet wide which gives storage room for wagons, buggies, etc., and at the back end there are watering arrangements and a depression in the cement floor with a hole in it to carry off the water. In this depression buggies and other rigs are washed and for this purpose it is a good plan to have either a force pump or a tank supply of water under pressure.

There is a feed door in the side of each box stall which hinges at the top and may be lifted to put grain into the manger boxes. Hay, however, is poked down from overhead through the hay chute. There are four double stalls and four single stalls which is a very good arrangement because some pairs of horses like to stand together while other horses are so quarrelsome that they must have separate stalls.

The feedway in front of the horses is narrow because it is not intended to store any feed on this floor, except in the corn-crib. There is a chute from the oat bin above with just a small box at the bottom to dip from. There is a lid to this box which shuts down in such a way that a horse could not lift it even if it should get loose and crowd into this narrow feedway.

The corn-crib portion is boarded tight on the inside, but the two outer sides are slatted. There are small doors outside near the top of the crib to shovel the corn through and there is no inside door opening into the crib except from the feedway in front of the horses.

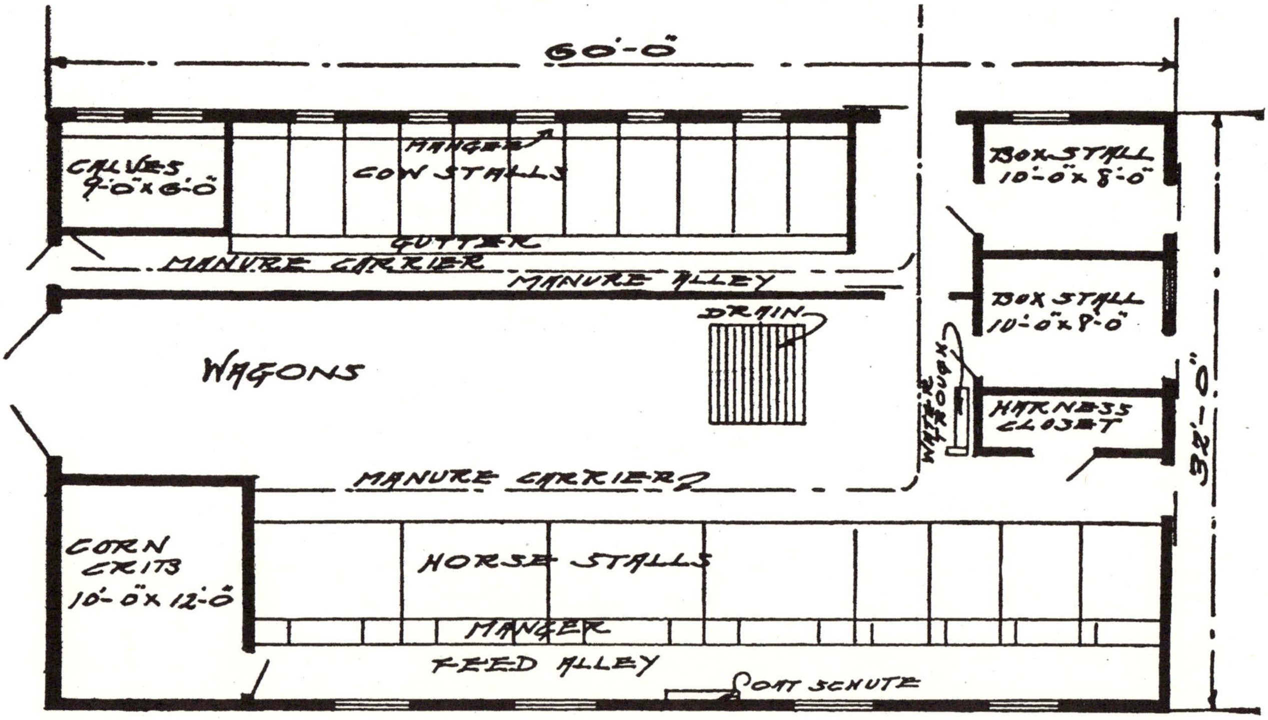

FLOOR PLAN

Hay is put in from the front of the barn by means of a hay fork and oats are lifted to the second floor by a sling drawn up with a horse, one, two or three sacks at a time.

This arrangement stands all the stock with their heads toward the walls, a different arrangement from most barns. When the feed is put down from overhead there really is very little objection to an arrangement of this kind in a small barn and there is an advantage so far as the horses are concerned in that they may be unhitched from the wagon, go at once to the watering trough and drink, then find their own stalls without assistance from anybody.

A long narrow harness closet with two doors makes a good place to hang harness so that there is no excuse for having it scattered around on pegs in the way.

HAY AND GRAIN BARN—A167

Cost of Blue Prints, $15.00

A long barn designed to hold a good deal of hay and grain is shown in this illustration. It is a timber frame covered with eight inch drop siding.

The track for the hay fork is suspended from the peak by seven-eighth inch iron rods and the track extends the whole length of the building and projects several feet at each end. This arrangement makes it convenient to fill the barn from either end or from both ends as occasion requires. There is a driveway crosswise through the barn at the center. This driveway is floored with a two inch plank floor, but it is not necessary to floor the other part of the building except with round poles to keep the hay and grain sheaves off the ground. Such a barn is intended

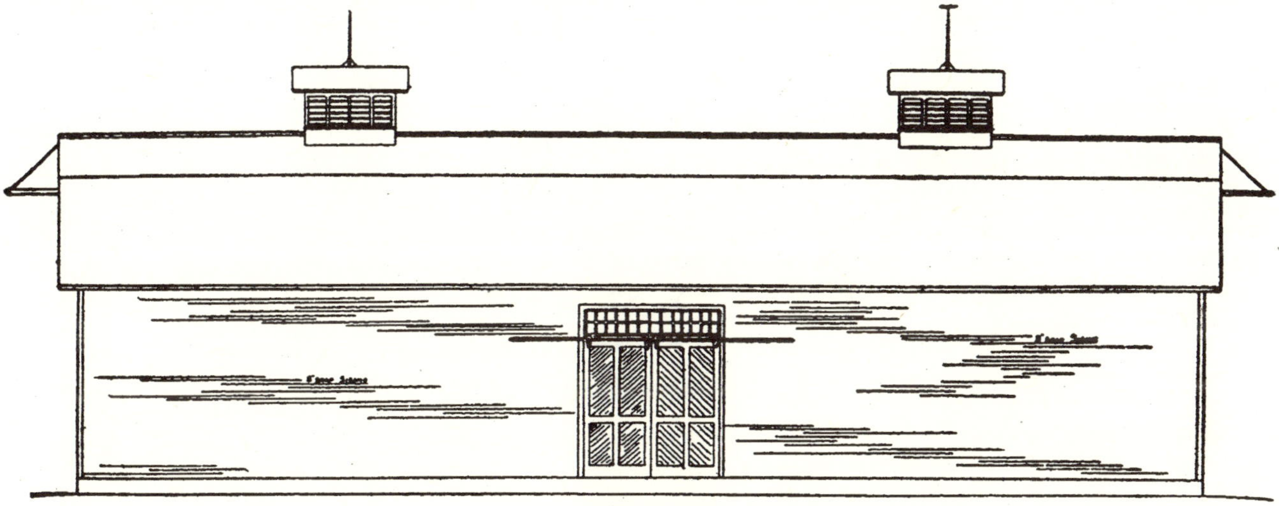

SIDE ELEVATION

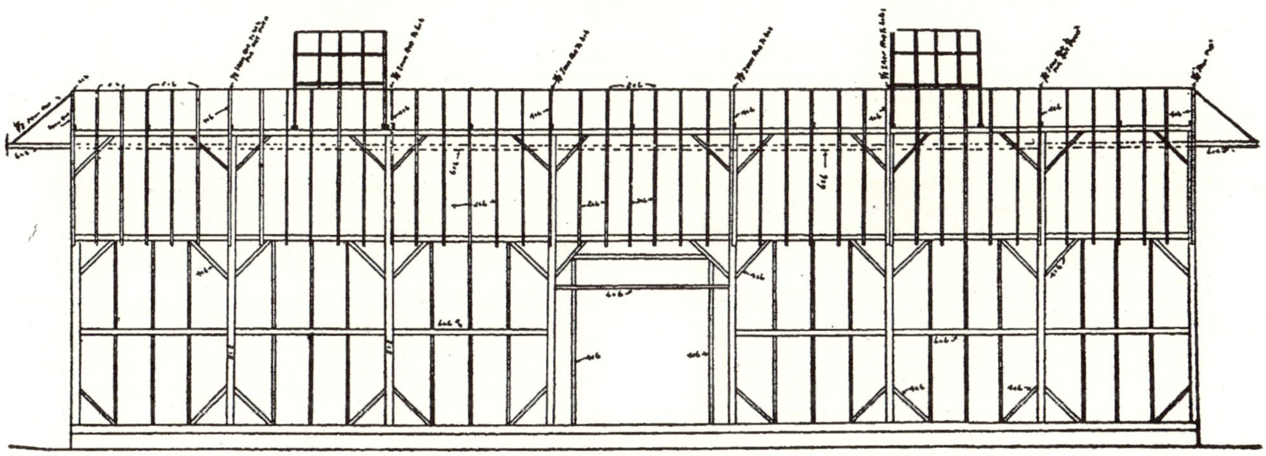

LONGITUDINAL SECTION

more for storage on large farms where considerable grain is harvested and hay cut either to feed or for sale.

It is not necessary to have such a barn near the other farm buildings. Many farmers prefer to have it convenient to the fields because it is never used for housing stock unless it be sheep and they don't require quite such frequent attention as the other animals.

The cross center floor is intended for threshing, but there is no provision for storing threshed grain. It is supposed that there is a granary near the house and other buildings and it is better to haul the grain from the machine.

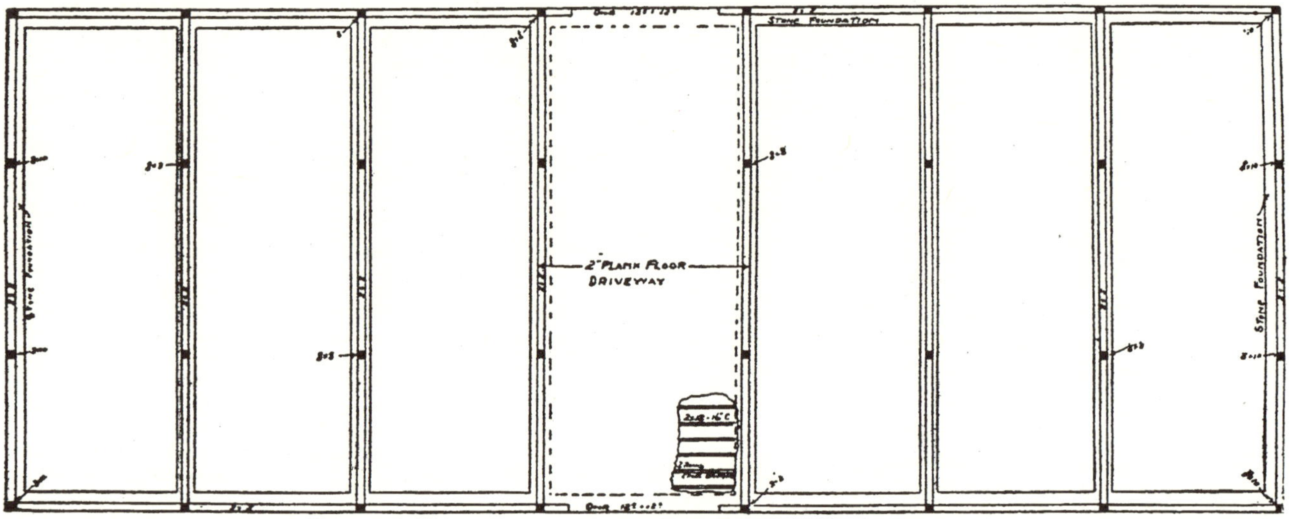

FLOOR PLAN

PRETENTIOUS STOCK BARN—A179

Cost of Blue Prints, $35.00

This pretentious stock barn is very complete and of an elastic pattern, so designed that its capacity can be increased by building on to the gable ends and extending them out any distance that may be required without affecting the general arrangement or exterior architectural proportions in the least. The two wings to the right and left of the silo contain the young stock and horses respectively and face the south. These two wings form a sort of court around the silo, admitting the sun, but obstructing the severe storms and giving shelter to the stock. The silo is well situated with reference to feeding, being in the middle of the cow barn. The cows stand back to back, which is of great advantage in cleaning out the gutters, as all the dirt can be handled from the center driveway and carried to the manure pits to the right. To the left hand or west end of the cow barn is a large room for imple-

SOUTH ELEVATION OF STOCK BARN

ments, wagons, harness cases and stairways to upper floor which contains grain bins, storage rooms for light machines vehicles, etc., and sufficient hay and feed room for all stock.

This building has a concrete foundation with the concrete walls extending about 2 feet above the cement floor level in the stock rooms. This prevents any moisture from getting to the framework and also makes a very sanitary and durable building. The frame walls are constructed of two by six studding covered with tar paper and drop siding on the outside and tar paper and matched sheathing on the inside.

The lower story has two rows of posts which support the upper floor and also serve to hold the stanchions and stall partitions. The upper story is of a single span, braced roof which allows the free use of a trolley hay fork the full length of the building.

The roof is of green stained shingles, of Dutch colonial architecture, and not only of a very appropriate design, but its shape adds greatly to the storage capacity of hay, grain, etc.

There is an embankment driveway on the north side which admits hay wagons into the upper floor for the unloading of hay, grain, etc. The silo is of frame construction lined on the inside with paving brick, making it absolutely air tight and almost frost proof. There is a trolley track feed carrier hung to the ceiling of the lower story, which simplifies the feeding. The building, as the cut shows it, will accommodate 100 head of cattle and nine horses.

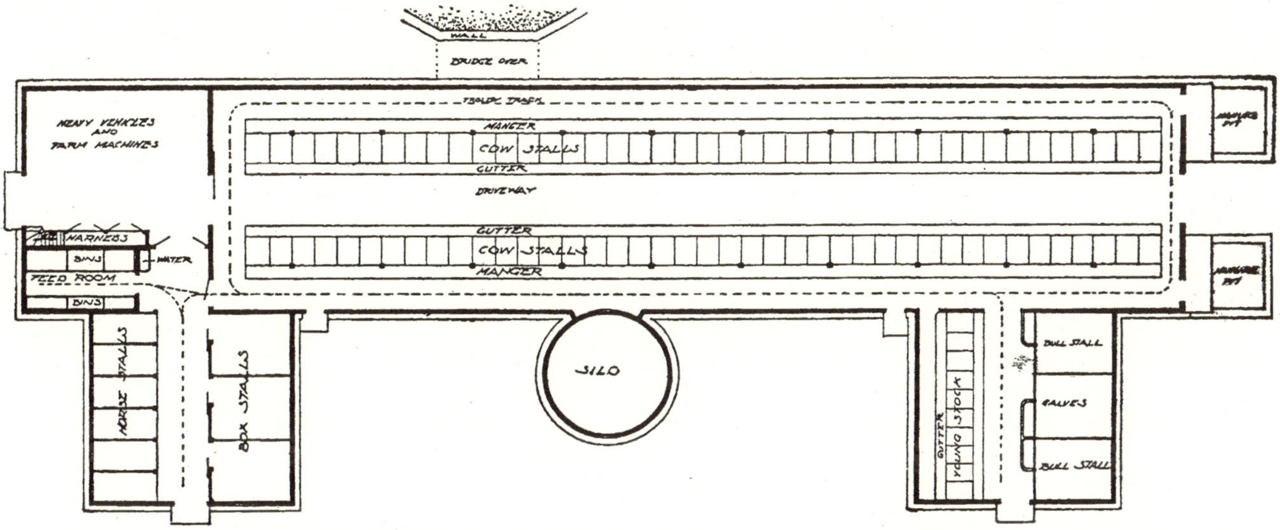

BARN NEAR ST. FRANCISVILLE, ILL.—A188

The accompanying is a medium large, plain and very serviceable barn on an eighty acre farm built at very low cost, the timber being furnished and most of the work being done by the farmer himself who owns a small tract of timber from which the logs were cut, furnishing all the lumber for the frame and siding.

The barn is of red oak lumber undressed and unpainted. The frame is what is known here as the "spiked" frame, three two by eight inch plank being spiked together, making finished timber six by eight inches. The barn is sixty by forty-eight feet with twelve foot driveway lengthwise through the center of main building and an inclosed twelve foot shed on the south stabling twelve dairy cows. There is a twelve foot open shed the entire width of the west end. It is twenty feet to the eaves of the main part, fourteen feet to the eaves of south cow shed and thirty-

five feet to the comb. The mow above covers the entire floor, sixty by forty-eight feet, and will hold eighty tons of hay. Hay is taken into the mow from outside east end from large door just under comb, and it has modern equipment of track and hayfork.

Horse stalls are arranged on either side of the center driveway, the horses facing the walls, and having capacity for sixteen horses.

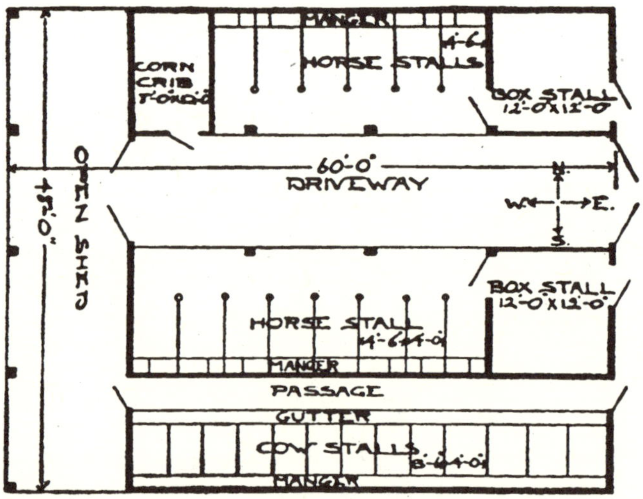

The barn is on a foundation of natural stone pillars with earth floor, and the building is constructed to fit the hill, which slopes to the west, east posts being shorter and west posts being longer. There is a small corn-crib in the northwest corner, and two box stalls at east end.

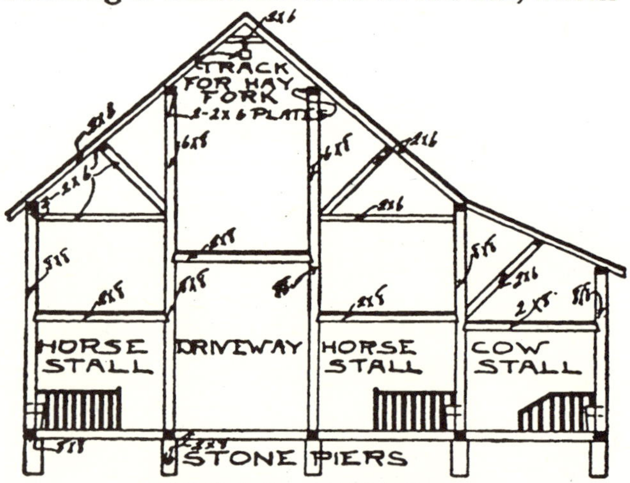

CROSS SECTION

WABASH COUNTY, ILL., BARN—A185

A medium large barn on the 160-acre farm of Mr. J. O. Wood, of Wabash Co., Ill., one of the most scientific farmers of the community. It is a plain structure with no sheds, and is ninety feet long by forty-six feet wide by twenty feet to the eaves and thirty-seven and one-half feet to the comb. The frame work consists of seven bents placed fifteen feet apart. There is a driveway through the center closed with

double hinged doors at each end. The large mow above holds ninety to one hundred tons of hay which is taken up inside at the end of the driveway, the floor being afterwards replaced and the section blown full of shredded fodder.

One side below is used for horses and the other for cattle with mangers adjacent to driveway. It will stable fifteen horses and from twenty to forty cattle.

Good ventilation is supplied by small doors on a level with the heads of the animals. The barn is built on solid brick foundation and the frame is of sawed oak timbers; siding is pine ship lap and roof red cedar shingles. It was built in 1903 at a cost of $1,500, Mr. Woods furnishing timber for frame from his own forest and doing all of his own hauling. The only change he would make in building again would be to build four feet higher for greater mow capacity.

This barn is unusually high at the sides. Boards twenty feet long for boarding up and down are not usually easy to obtain and this must be taken into consideration when planning to build. Some farmers when depending on local carpenters prefer to make a building higher at the sides with a plain straight roof like this rather than to undertake a curb roof with the

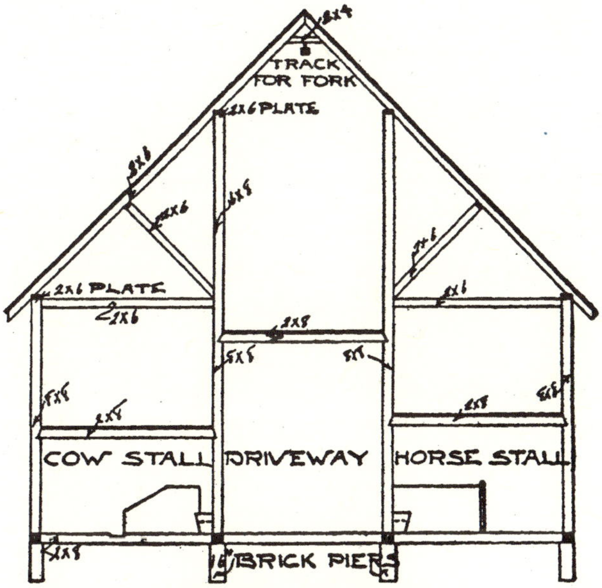

CROSS SECTION

extra skill and care in framing that such a roof entails.

We show this barn as an old fashioned

type that has a great many advocates among practical farmers. There are restrictions in barn building as well as other things. What suits one community or one farmer is not to be recommended for another and a great deal depends on the carpenters within reach. Almost any local carpenter can lay out a barn like this, take the usual handy men about the place as helpers and push the work along from beginning to finish without a hitch; when the same carpenter with the same help would be bothered to death with the intricacies of a more complicated building.

There are more economical barns than this in regard to space because you lose a good deal out of the center of the barn with such a long driveway. On the other hand one half of the barn is devoted to cows and this driveway answers for a feed room to drive through with a hay rack loaded with roughage from the fields for both cows and horses and the driveway is supposed to be floored over with timbers heavy enough to support a mow above.

Openings are shown by the dotted lines for putting down hay and straw in winter and there is another opening in the center for the hay fork in summer when the barn is being filled. This center opening should be covered with poles or planks and hay thrown over it to prevent a draught. The hay chutes should be boxed around and closed at the floor level with weighted trap doors for the same reason. One of the greatest objections to openings of this kind is the draught they create. Ventilation is absolutely necessary where a number of animals are kept together, but ventilation does not mean a draught.

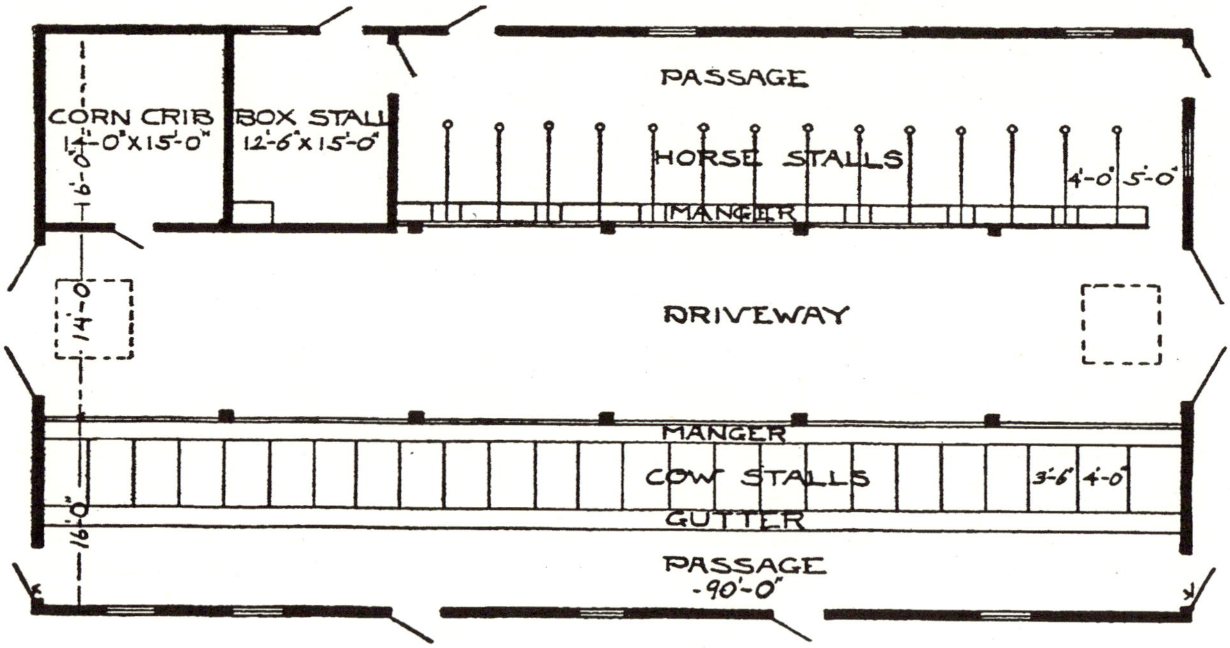

YANKEE BARN—A134

Cost of Blue Prints, $5.00

A style of barn that is often seen in New England is given in plan (A134). The horses and cows occupy part of the first floor, leaving a space in one corner that makes a convenient storage for farm tools. There is a driveway through this part of the barn and the door is large enough to get in with a hay-rack or a grain drill.

The upper part of the barn is used almost altogether for hay storage, the hay

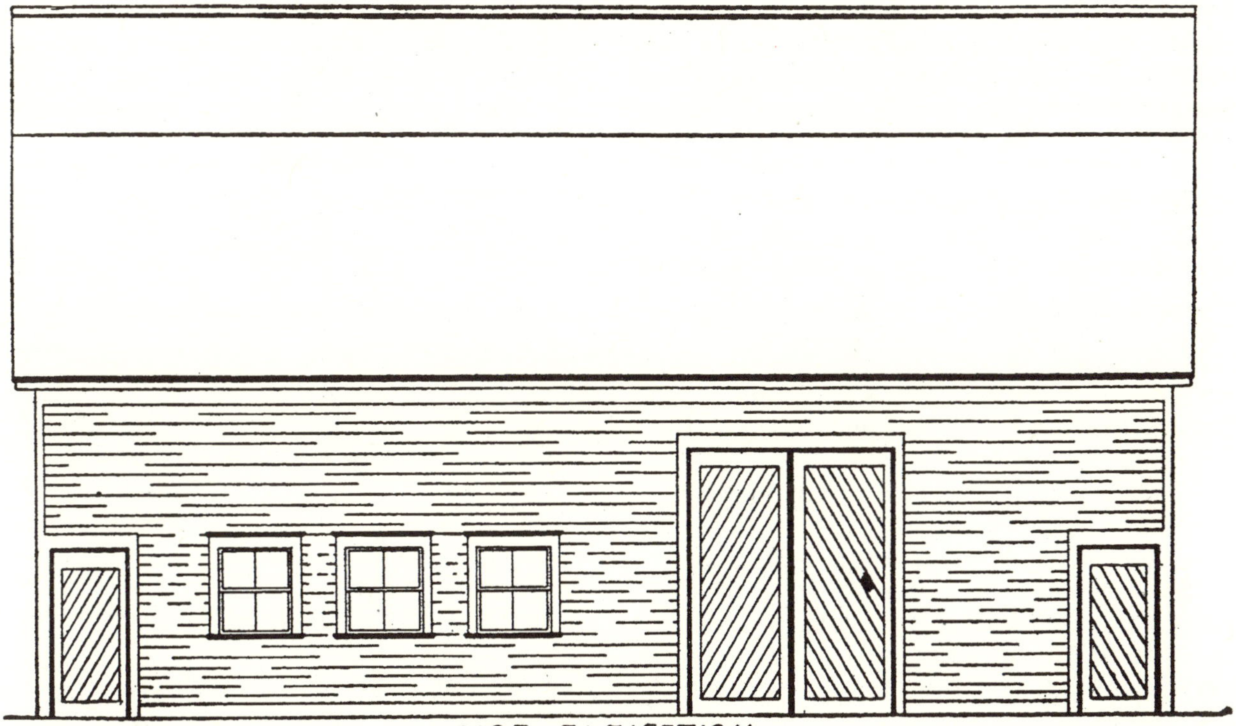

SIDE ELEVATION

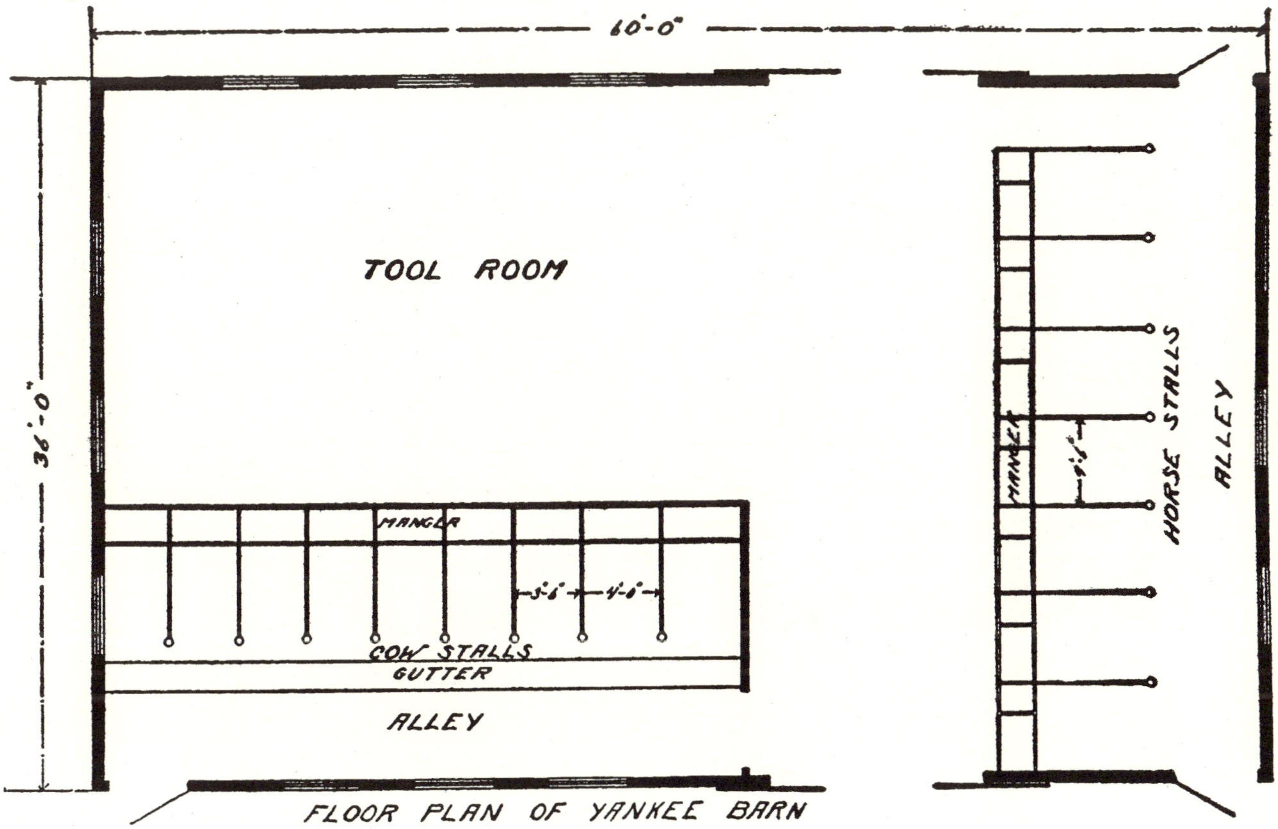

FLOOR PLAN OF YANKEE BARN

being lifted from the driveway by a horse-fork. It makes the stable much warmer to run the partitions in front of the cows and horses to the ceiling above. Unfortunately, too many farmers are careless about such things and their animals often suffer in large draughty stalls.

This barn is thirty-six feet wide by sixty feet long, not very large on the ground for a farm barn, but the shape of the roof helps out very much in storage.

It is floored over with the exception of an opening over the driveway and as this floor is only nine feet above the ground it leaves a very large loft.

There are a great many Yankee barns without so many windows, but the windows are a great advantage. It is much easier and more pleasant to do work in a light barn and the animals do better. It is difficult to account for so many dark barns, except that the fashion was established when the country was new and window lights were a great deal more expensive than they are now. Glass and sash are just about as cheap as siding, there is no economy in building dark barns.

BARN FOR A SMALL FARM—A160

Cost of Blue Prints, $10.00

This is a small barn for a small farm where four or five horses are kept besides a few milch cows and a little other stock. This barn was designed for 10 cows, five horses and about fifty fowls and there is room for a couple of breeding sows. In every stable a box stall or two comes in handy. A box stall is almost an absolute

necessity sometime during the year either for sick animals or because some special attention is required.

The entire upper part is floored and there is an opening over the storage and implement room to pitch up hay, straw and other forage. This same opening answers for passing feed down to the mangers from the feed lofts.

There are windows all around this barn for light and ventilation; a provision that is too often left out when farm barn plans are made. It is not necessary to shut a barn all up dark, and it is not advisable to do so. Windows do not cost much more than siding and the sun and light let in is a great advantage to stock.

The floor of this stable should be of concrete with the upper layer an inch thick composed of one part Portland cement and two parts clear soft sand but in making a floor like this is should be remembered that hard smooth cement is slippery and dangerous. The passage way may be marked off in diamonds with a regular tool which presses into the soft cement about one-half inch deep, but if the work is done on the farm and the usual mason's implements are not at hand, a smooth rake handle may be used by imbedding it in the soft cement half its thickness. The handle should not be more than three-fourths or seven-eighths of an inch in diameter.

Unless the concrete foundation in this

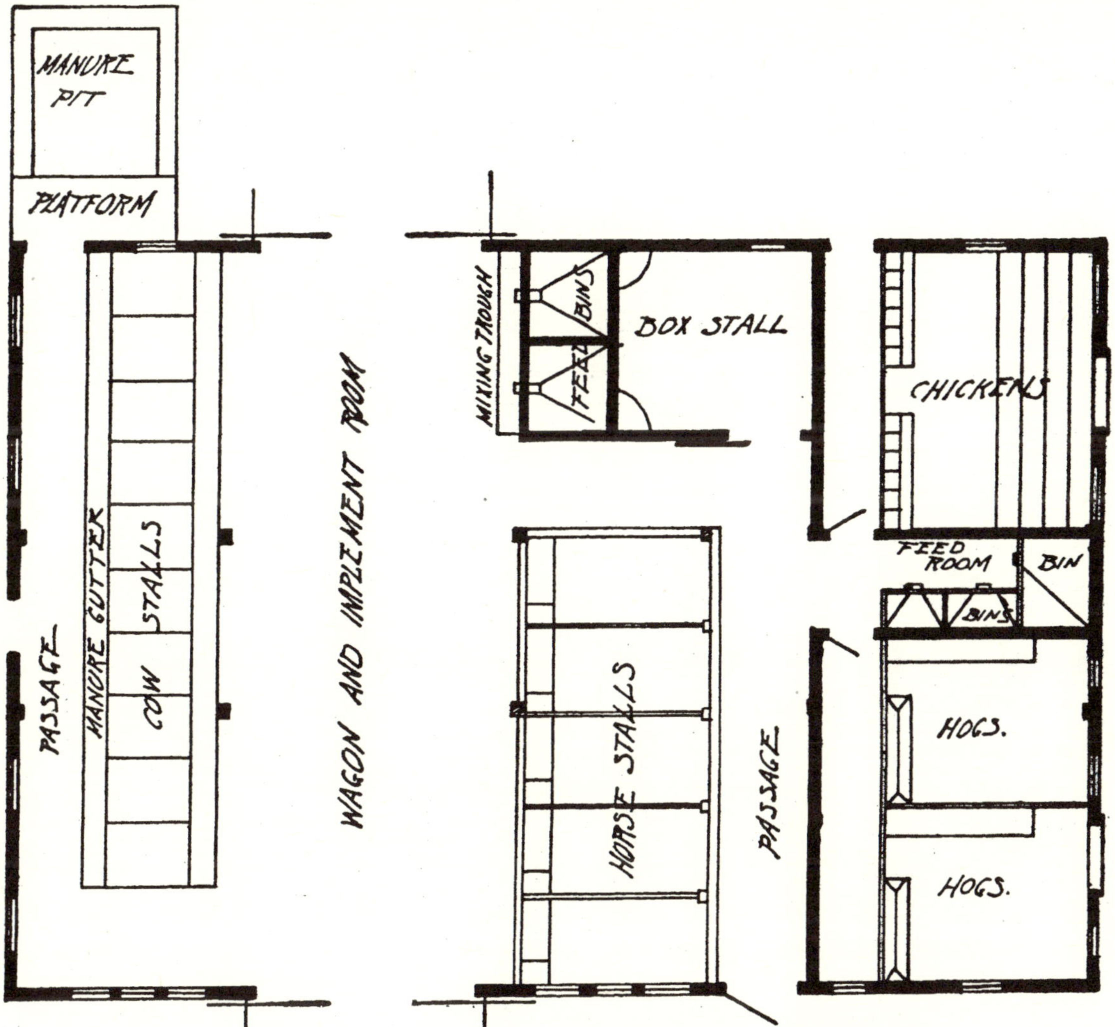

PLAN OF BARN

driveway is of superior quality the cement top layer should be more than an inch in thickness, perhaps two inches in the center, tapering to an inch at the sides next to the stalls.

In laying a concrete floor in any building it is necessary to run a wall around the outside and this wall should extend below frost. If the ground is inclined to dampness, it is better to run a three inch or four inch drain tile all around the wall along the bottom and the outlet of this tile should be carried away from the building eight or ten feet and terminate in a drain.

SMALL FARM BARN—A169

Cost of Blue Prints, $10.00

A neat little barn that is well proportioned and suitable for a farm of twenty or thirty acres is given in these illustrations. There is a threshing floor in the middle with wide double doors in the north side as well as in the south side making a good liberal passageway through the center of the barn.

stalls. The cow stable side has a ceiling seven feet high. Cows don't get their heads up as high as horses do and they don't need such a high ceiling. Cows keep warmer in a stable with a low ceiling and if there is plenty of chance for the air to get in and out again they have good ventilation.

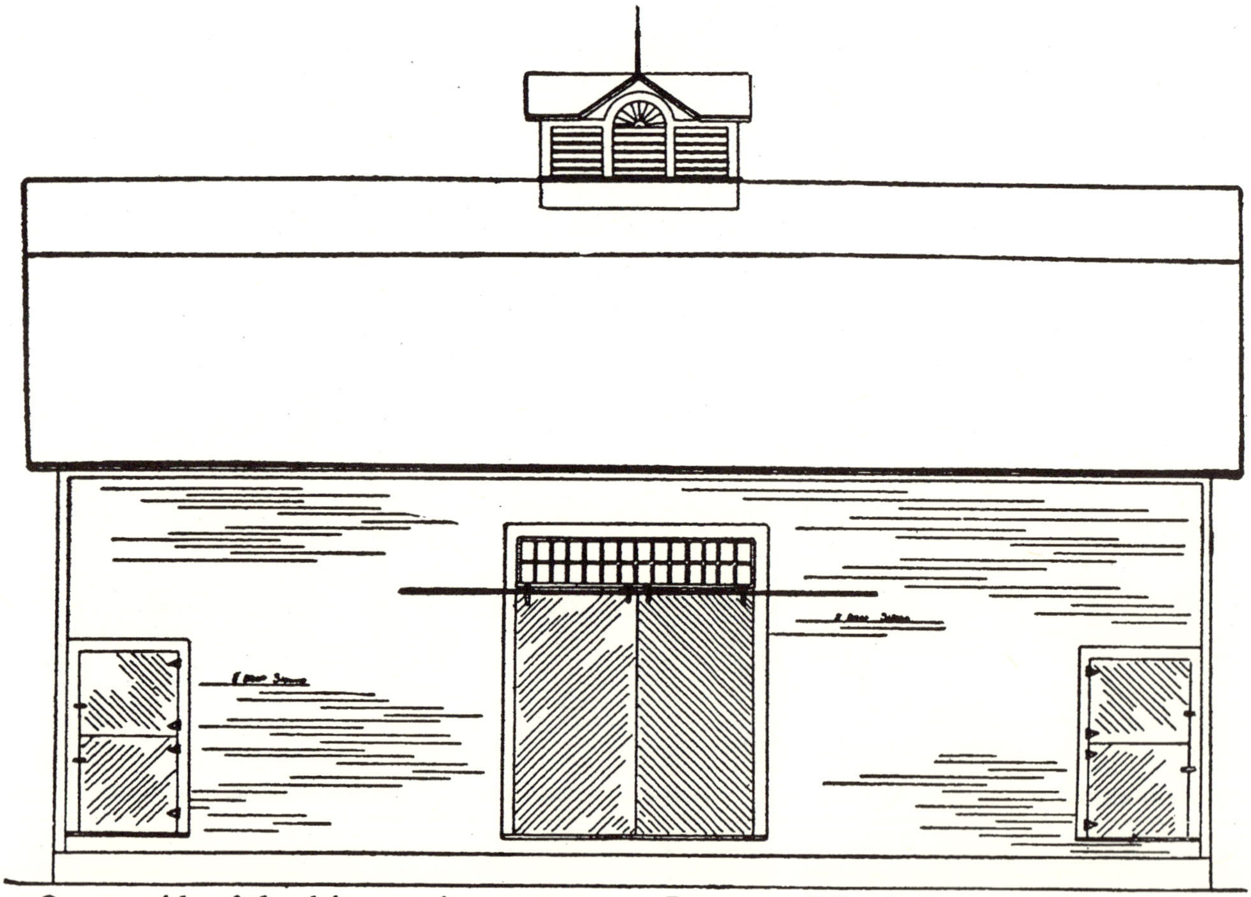

On one side of the driveway is a granary and stabling for three horses with a nine foot ceiling. A third of the barn on the other side of the driveway is made into a cow stable making seven good roomy

It seems difficult for some live stock men to understand this phenomenon. The reason is the air circulates more freely when it is warm. The body heat of seven cows in this stable with a low ceiling will warm

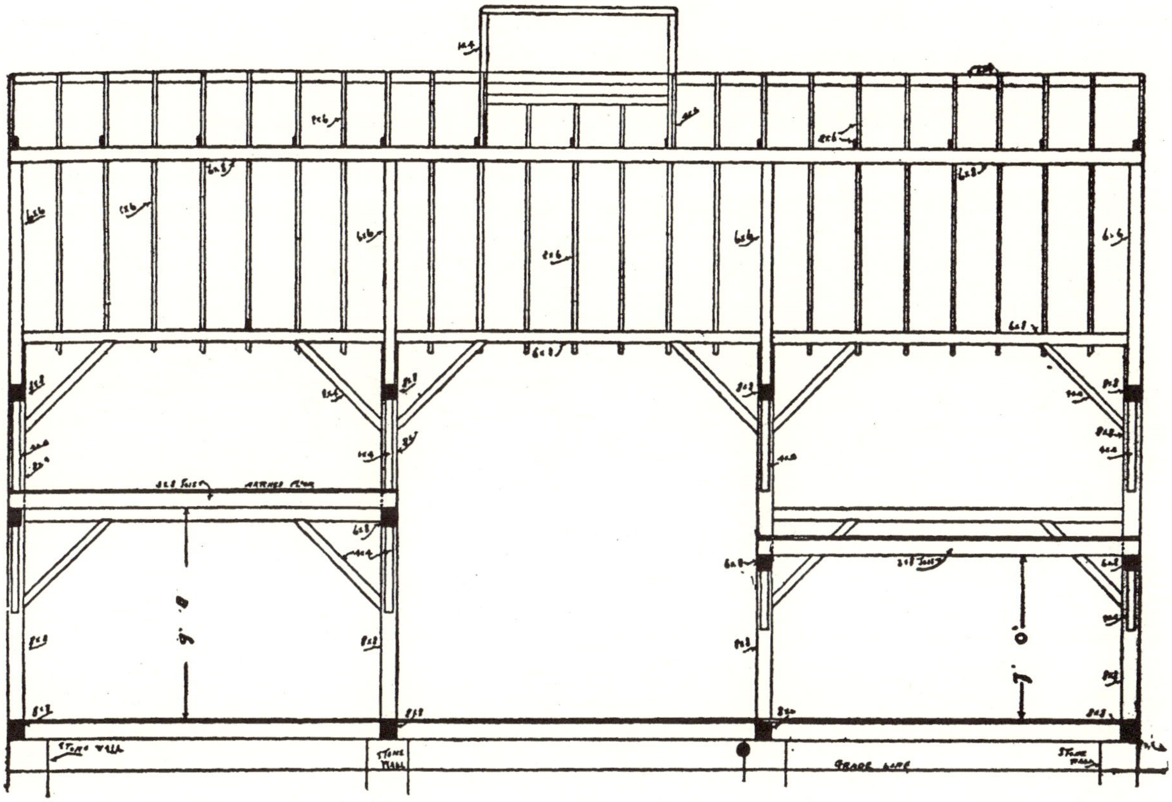
9' 8
7' 0"
Stone Wall
Grade Line

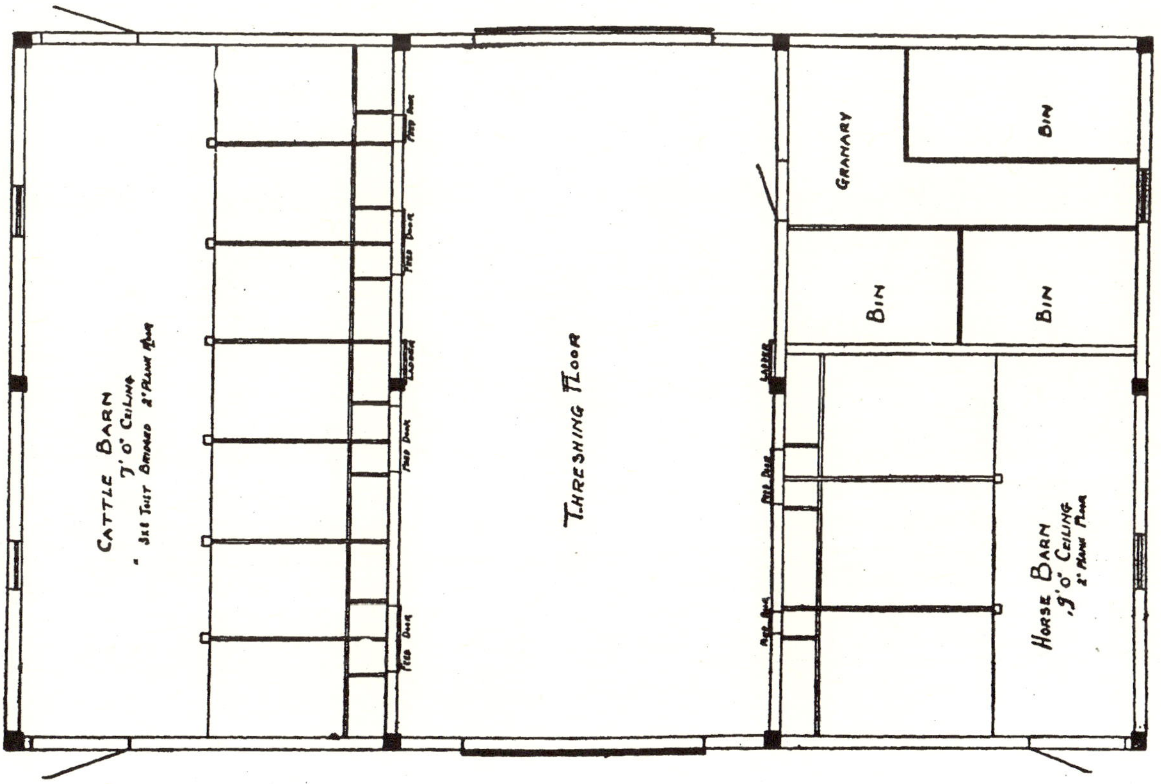
Cattle Barn
7' 0" Ceiling
3x8 Joist Bridged 2" Plank Floor
Threshing Floor
Granary
Bin
Bin
Bin
Ladder
Feed Door
Horse Barn
9' 0" Ceiling
2" Plank Floor

the air sufficiently to keep it in circulation. If there are openings where the fresh air can get in, the foul air will find its way out and there will be a constant change.

Both the cow stable and horse stable are boarded up in front, but barn boarding usually is not very tight. Unless matched stuff is used there is a little opening between the boards that allow for the escape of a good deal of bad air. There usually is considerable space around the doors. There are feed doors in front of the stable so the fodder may be put in from the barn floor.

It is hardly necessary to use a horse fork in a barn of this size. The flooring overhead does not cover the whole of the threshing floor so that hay and grain in the sheaf is forked up by hand. It will be noticed by referring to the transverse and longitudinal sections that the timber is very carefully planned for size and length in proportion to the building. Every stick is necessary but there is not a piece too many.

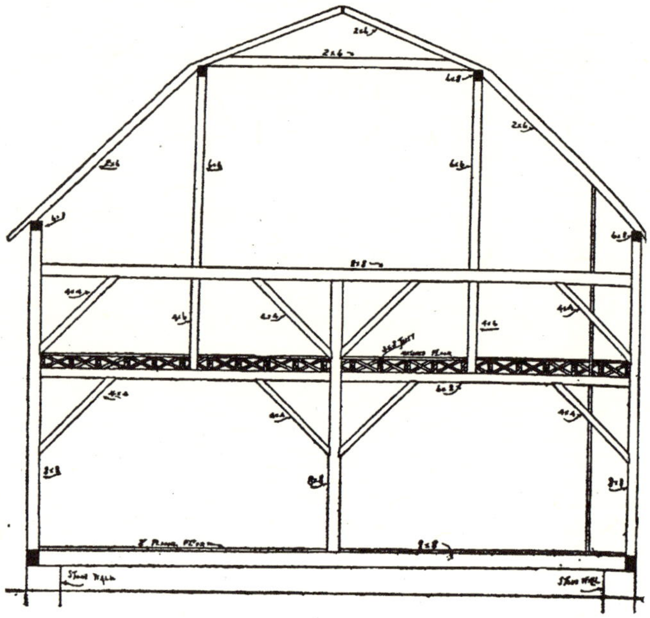

KESLER BARN—A189

The illustration shows a very attractive and handy barn. It consists of main part twenty-eight feet wide by fifty-two feet long, eighteen feet to the eaves and thirty-

five feet to the comb. On the north is a low sheep shed forty feet long by fourteen feet wide, and extending east of this is another low addition twenty-six feet long by eighteen feet wide, the north half of which is used for horse stalls, hog house and corn crib, the corn crib being built over the hog compartment. The south half of this extension is an open shed. Also an open shed extends along the east side of the main building and is equipped with mangers for six horses to be used when extra teams are at work on the farm, as at threshing time.

A driveway runs through the main building from east to west, on one side below which are horse stalls and corn crib. On either side of the driveway above are hay mows which hold about fifty tons of hay which is taken up from the center inside.

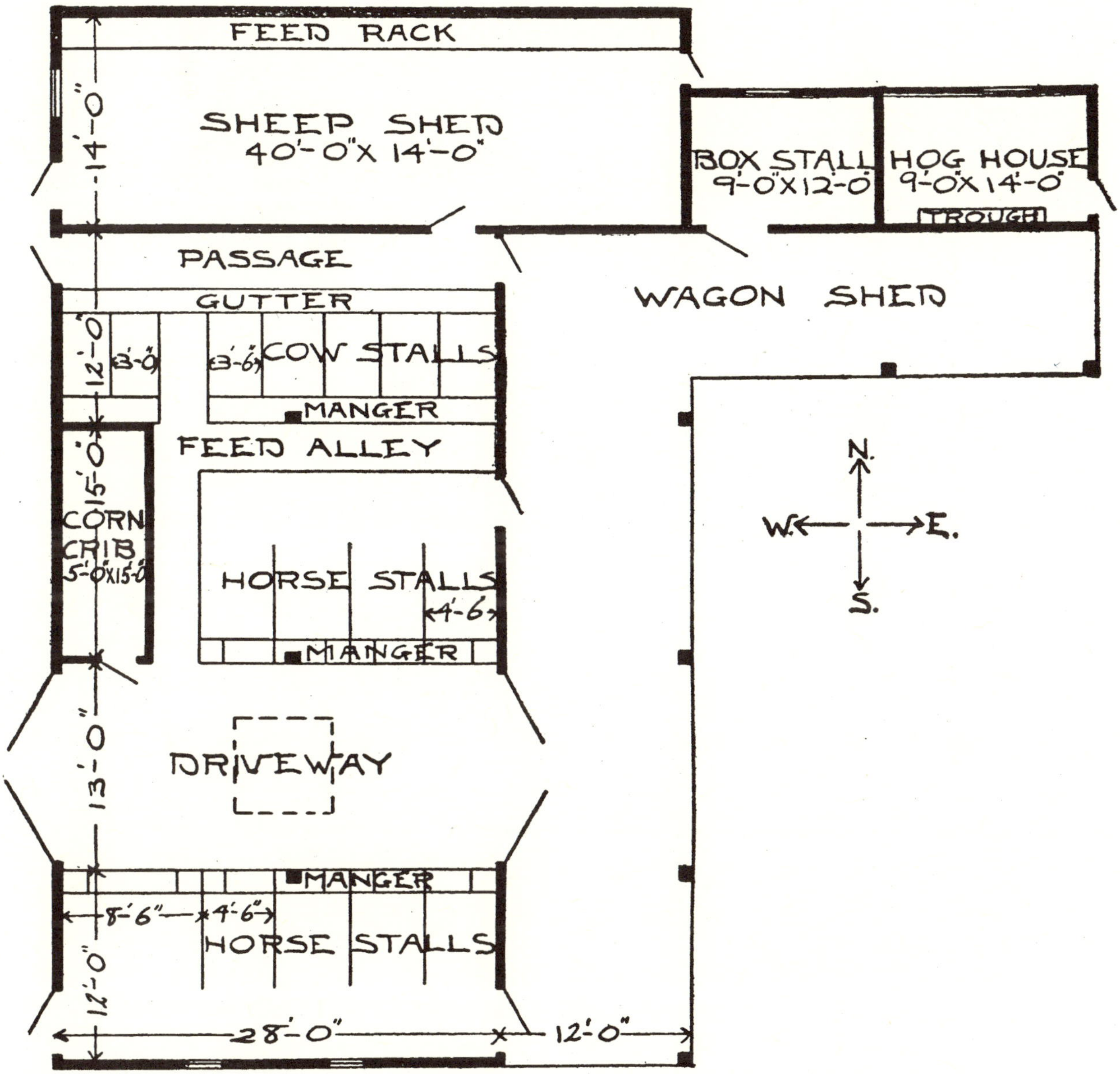

North of the middle horse stalls and crib are the cow stalls, a three foot feedway running between them.

The plan as given includes a leanto on

the north end which extends all the way across the end of the barn and is attached to the open shed which contains a box stall and a hog house and joins the northeast corner of the main barn. The shed for sheep is a good arrangement and it is plac-

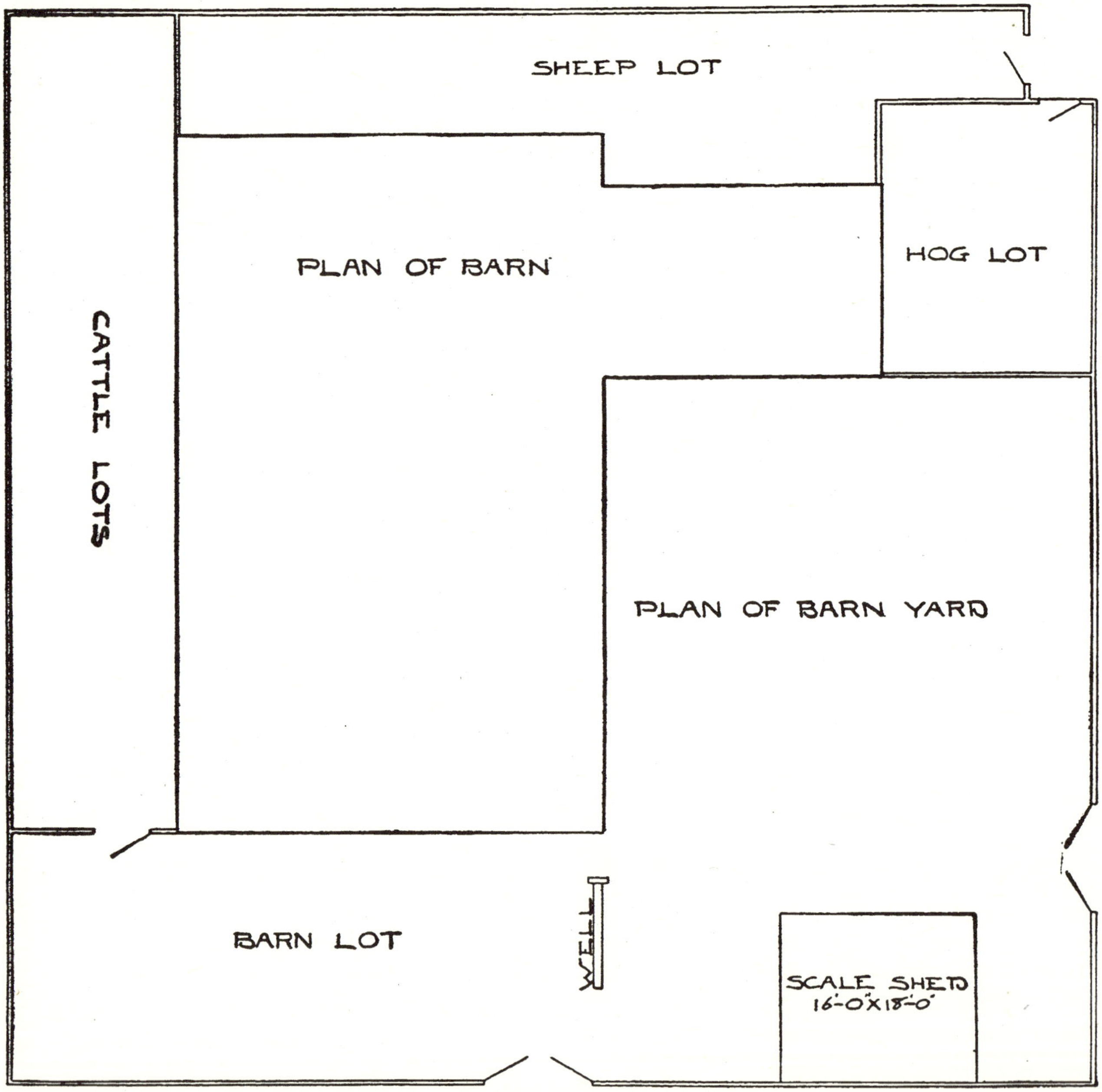

ed right. Sheep have no business in the common barnyard; they are likely to get hurt and they are a nuisance at feeding time. Sheep carry their own blanket with them and they require no warmth from the building except at yeaning time. A shed like this with a roof sloping to the north makes a very good shelter.

In this case the feed rack is built on the north side of the shed and it may be filled by putting down hay from the mow overhead. The mow is not very large, but as the shed is only fourteen feet wide feed is easily shoved through from the main barn. This method of feeding is much better than putting hay down through an open chute amongst the sheep. They crowd under the falling feed and they get their wool so filled with chaff and seeds that it affects the sale of it, besides the dirt and dust is no benefit to the sheep.

Keep sheep dry and have a lot for them to run out in during the day time. Drainage must be looked to or water will accumulate when the snow melts towards spring.

As a usual thing it is not a good plan to have a hog house in connection with the barn but on a great many farms only a few hogs are kept and they are allowed to run out on pasture most of the time. In such cases a hog house built into the far end of this northeast shed is permissible, though not advisable. During these days of specialties it is better to have small movable hog houses than to let them come anywhere near the barn.

The box stall under this shed will be found very useful to stable a horse or cow when they need veterinary attention, or to hold a mare and colt, or for two or three spring colts during their first winter. In fact it will be a better plan to build two or three box stalls under this shed and make provision elsewhere for the hogs.

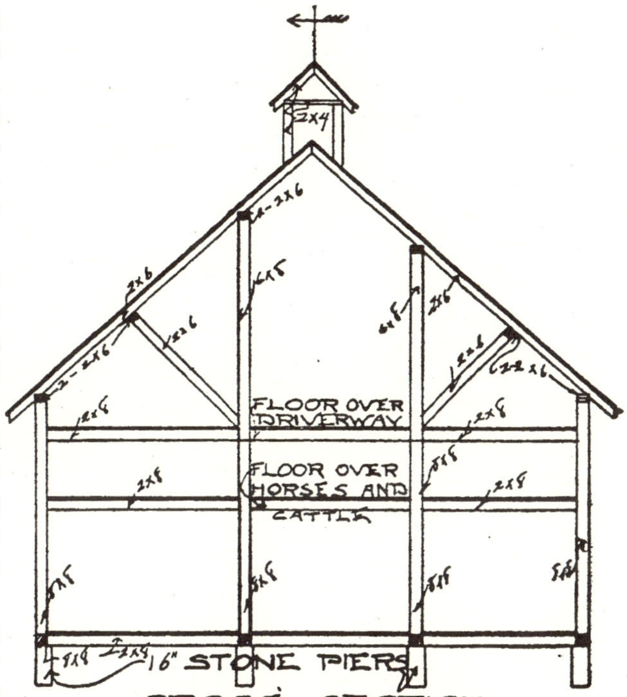

CROSS SECTION

BARN WITH ELL SHED—A163

Cost of Blue Prints, $10.00

A small barn with an ell shed attached is shown in this design. The barn proper which is 28 feet wide by 56 feet long is intended to accommodate five head of horses in about one-third of the floor space leaving the other part for a driveway with storage for grain, hay and farming implements. The whole of the second floor is given over to storage for hay or grain in the sheaf.

A hay bay extends from the ground to the roof in one end of the building but a floor extends over the stable and the greater part of the threshing floor. The threshing floor section may be partitioned off from the horse stable to make the stable warmer.

The shed forms an L running across the north and west sides of the barn yard, leaving the south side open to the sun. This arrangement breaks the north and the west wind and provides a comfortable barn yard for winter.

Stalls for 12 cows are built in the north

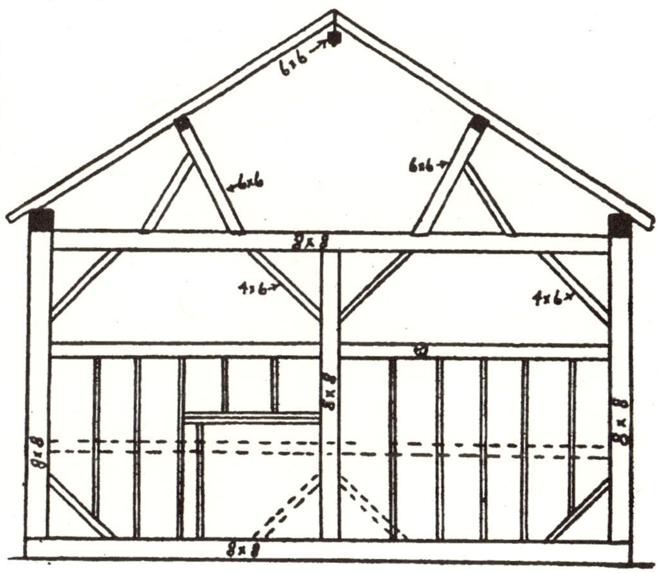

shed by putting two cows in each stall. This shed has a cement floor built like a sidewalk and the floor extends out under the projecting roof which comes over a few feet into the yard forming a protection against rain and snow.

It is something that every barn yard

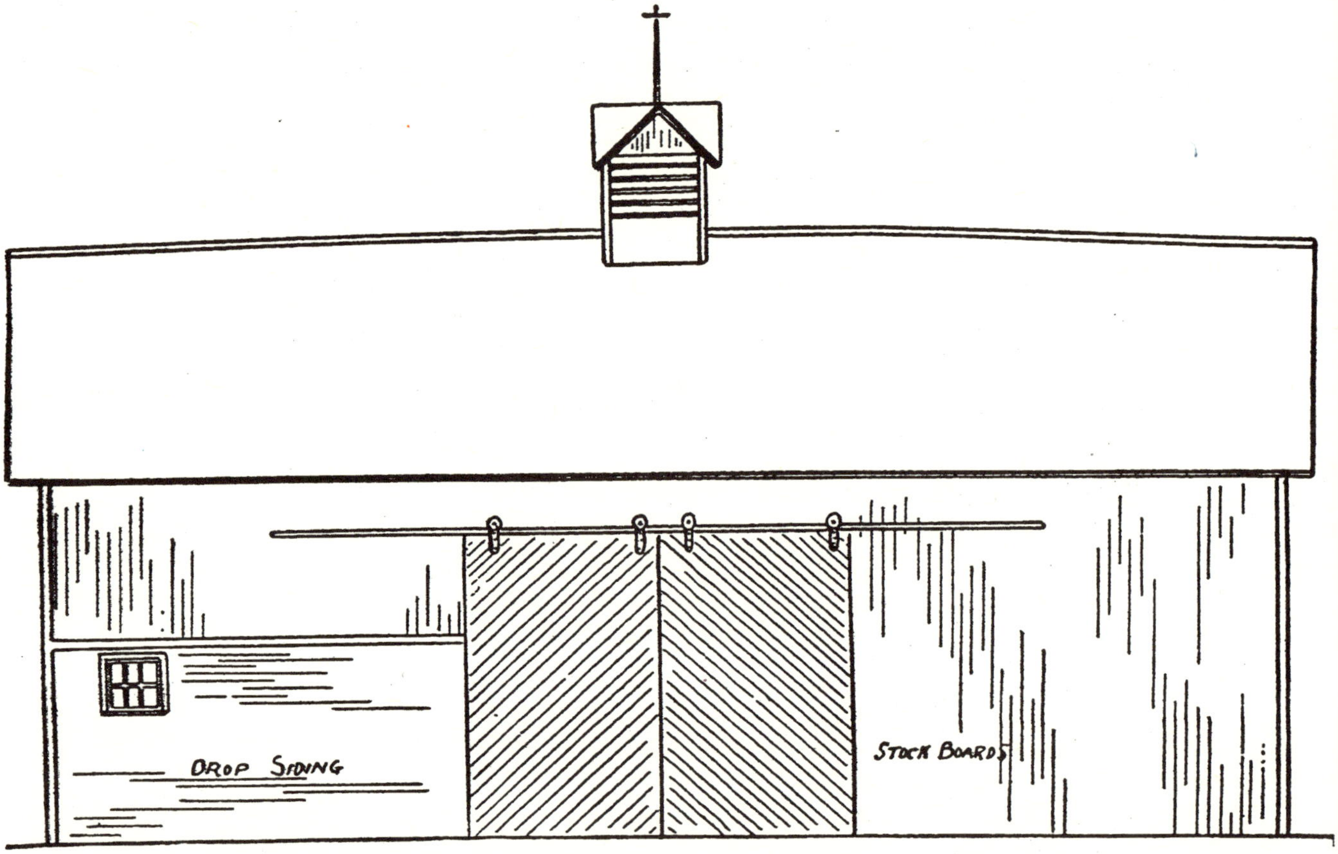

should have because there are times when the yard is wet and muddy in spite of ev-

ery precaution. A wide roof dripping into a barn yard is objectionable but the drip from this little short roof is insignificant.

The other part of the shed is open to the weather on the east side looking towards the barn, an arrangement that makes about as comfortable a barn yard as possible to obtain without roofing the whole thing.

This little barn with shed attachment is not expensive but is more convenient than some larger structures. The cost is within the reach of any farmer although he may not have more than 20 acres of land. A transverse section is shown giving a good idea of how the building is framed. It is a strong frame that is easily put together and there is no waste of timber.

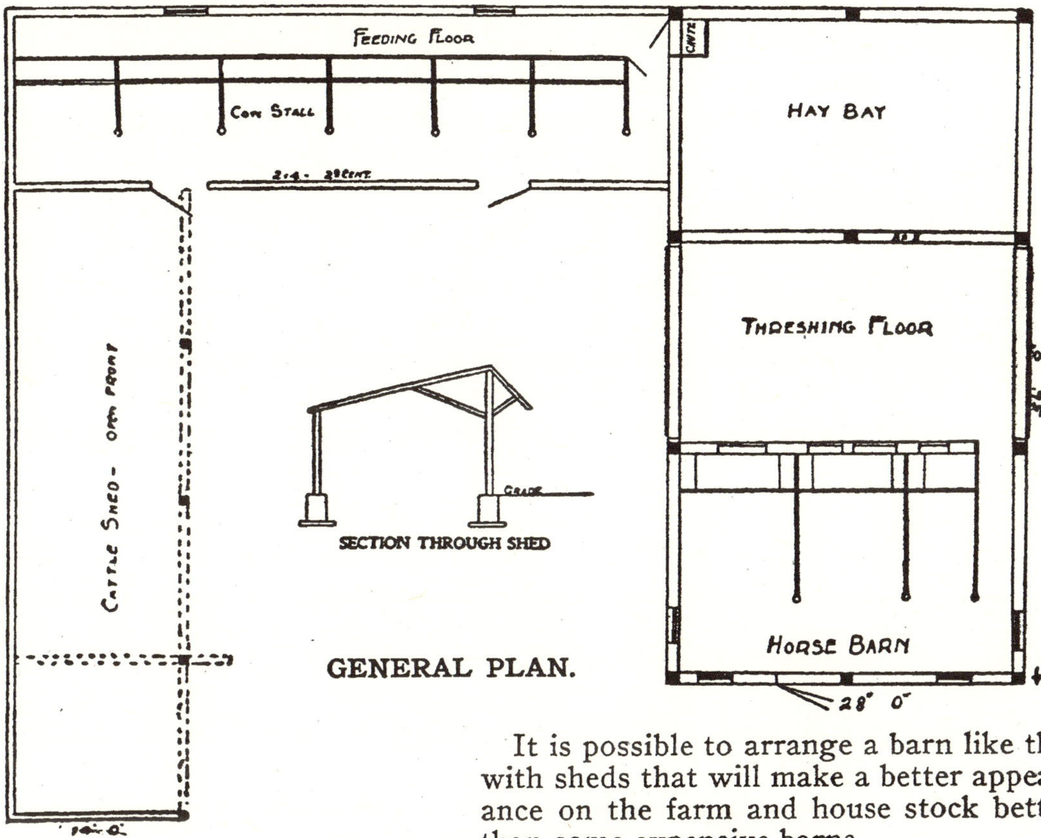

GENERAL PLAN.

It is possible to arrange a barn like this with sheds that will make a better appearance on the farm and house stock better than some expensive barns.

CATTLE BARN—A115

Cost of Blue Prints, $10.00

A medium sized barn to accommodate eight cows and six horses is given in plan (A115). The size on the ground is thirty-two by forty-four, which is not very large for a farm barn, but it is not intended to be a large one. The first floor is divided into three parts; the horses occupy one, the cows another and the middle section, fifteen feet wide, is left for general purposes. It answers for a feed room, storage for a wagon or two and general barn purposes. The second floor covers the whole building with a couple of hay chutes to let down feed and straw to the horses and cattle. It hardly pays to work a horse fork in a barn of this size. The stuff may be put in by hand from the outside through doors that open down to the floor. There is no waste space in this plan, every foot is made use of to the best advantage, and the barn will be found very useful on farms

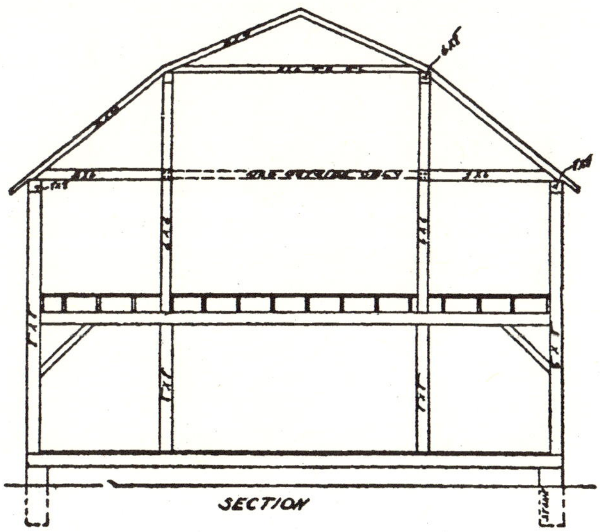

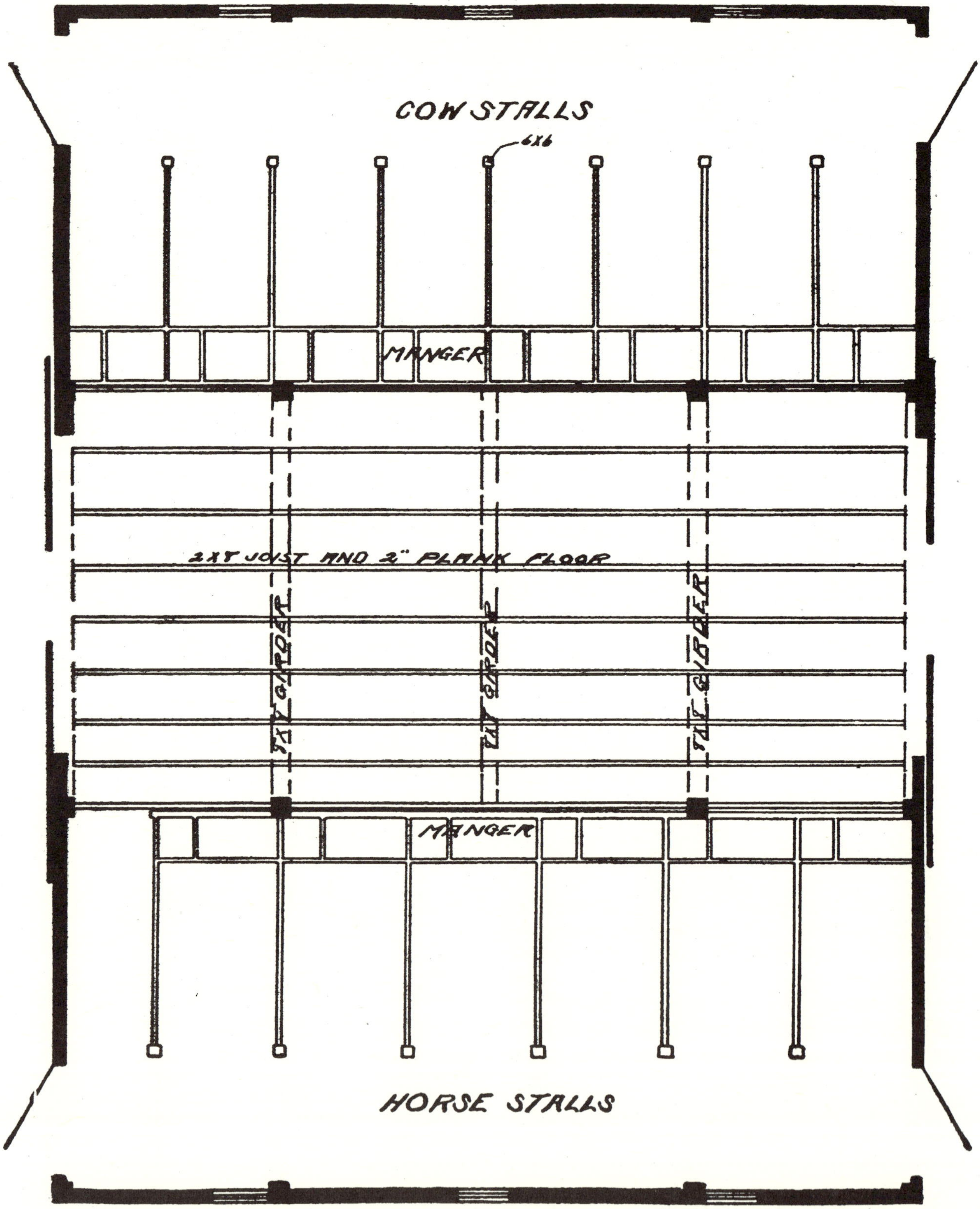

FLOOR PLAN OF HORSE AND CATTLE BARN

where a small number of cows and about the usual number of horses are kept. The plans show the construction in detail. It may be boarded up and down or covered with siding. May be made any length.

A good feature about this barn is that it can be added to without interfering with the general arrangement in any way.

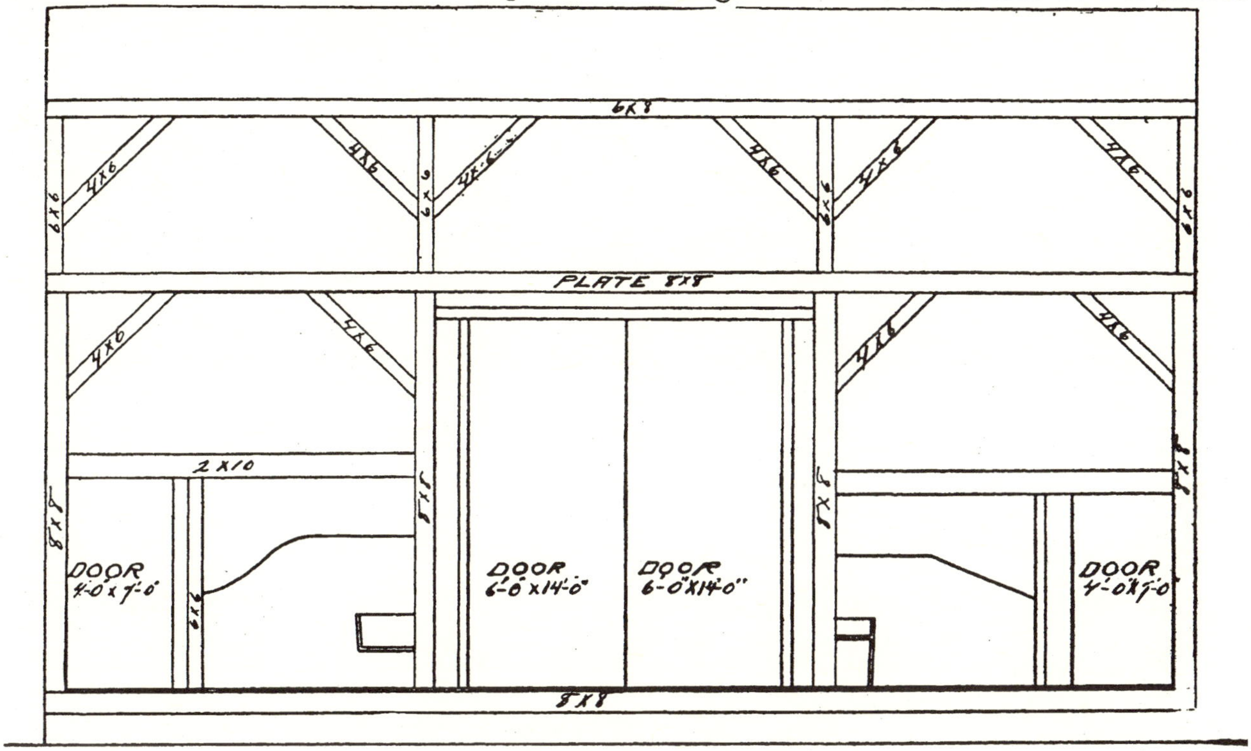

LONGITUDINAL SECTION

MOUNT CARMEL BARN—A186

This is a medium large, attractive and serviceable barn on a 120-acre farm. It stands on a solid concrete foundation wall three feet deep and eight inches wide at the top. It is eighty-four feet long, forty feet wide, twenty feet to the eaves, and thirty-eight feet to the comb. There are six bents of fourteen feet span each. The frame timbers are six by eight inches, oak stuff, and the rafters two by five inches of the same material. The siding is matched white pine painted red, and the roof is of red cedar shingles. There is a vegetable and fruit cellar under north end of driveway twelve by twenty feet by seven feet deep. The hay mow of the barn covers the entire upper floor forty by eight-four feet by twelve feet to top of side posts, and will hold about one hundred tons of hay.

It is fitted up with modern hay fork and track, and hay is taken into the mow from

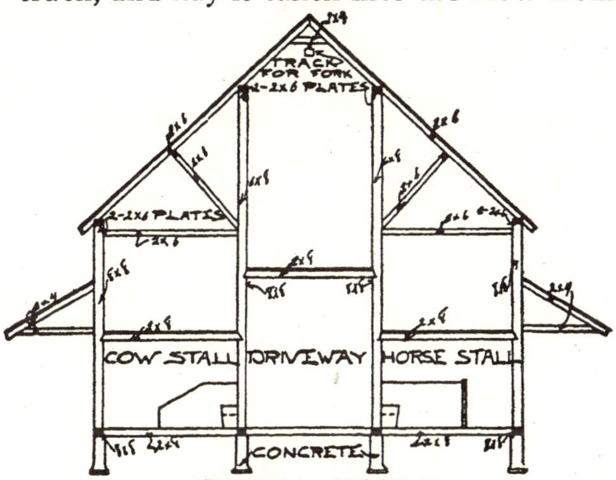

CROSS SECTION

either end, the openings being provided with pairs of swinging doors. Corn-cribs

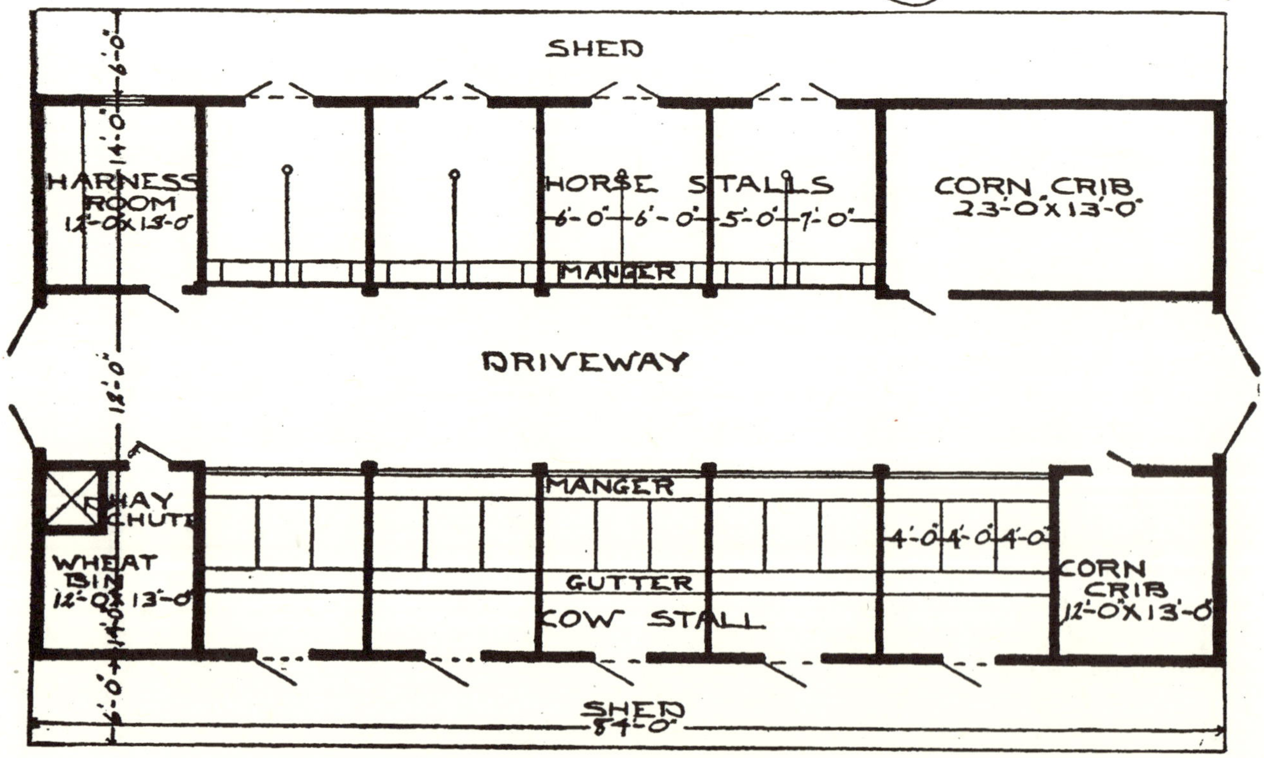
SHED
HARNESS ROOM 12'-0"X13'-0"
HORSE STALLS
6'-0" 6'-0" 5'-0" 7'-0"
MANGER
CORN CRIB 23'-0"X13'-0"
DRIVEWAY
HAY CHUTE
WHEAT BIN 12'-0"X13'-0"
MANGER
4'-0" 4'-0" 4'-0"
GUTTER
COW STALL
CORN CRIB 12'-0"X13'-0"
SHED
84'-0"
6'-0"
14'-0"
18'-0"
14'-0"
6'-0"

are built in two corners as shown and they are properly ventilated.

Running full length of both sides of the barn are self-supporting six foot sheds which allows full ventilation of the stables during summer through the open doors, but protection against both sun and rain. The barn will stable eight horses and twelve cows, the mangers all facing the long feed or driveway. The general appearance is imposing. The cost is $1,800, Mr. Risley doing his own hauling and furnishing the frame timbers from his own woods.

CONCRETE AND WOOD BARN—A213

Cost of Blue Prints, $15.00

Design of a practical farm barn suitable for a farm of eighty acres where the farm- The basement also contains a room for wagons and buggies with a wash floor de-

SIDE ELEVATION

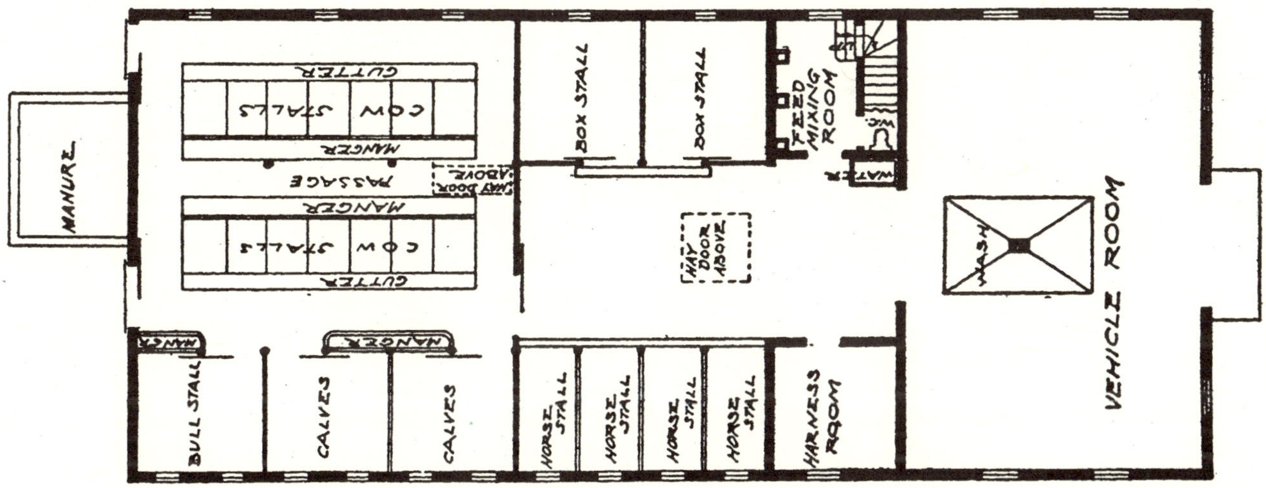

GROUND FLOOR

er wishes to keep a dozen cows and five or six work horses in the basement with storage overhead for hay and other roughage. pression in the center, an arrangement that makes a combination horse and cow barn with three box stalls for calves or other stock. It is a convenient arrangement where a farmer has only the one main barn.

Among the novel features in this barn is a cement floor with cement partitions between horse stalls. These partitions are made of reinforced concrete and are rounded down to meet the floor at the bottom which makes a very solid and lasting stall.

END ELEVATION

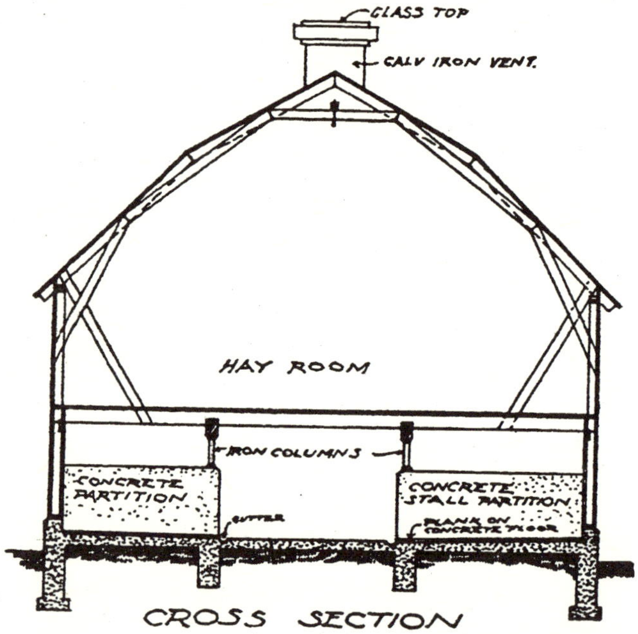

CROSS SECTION

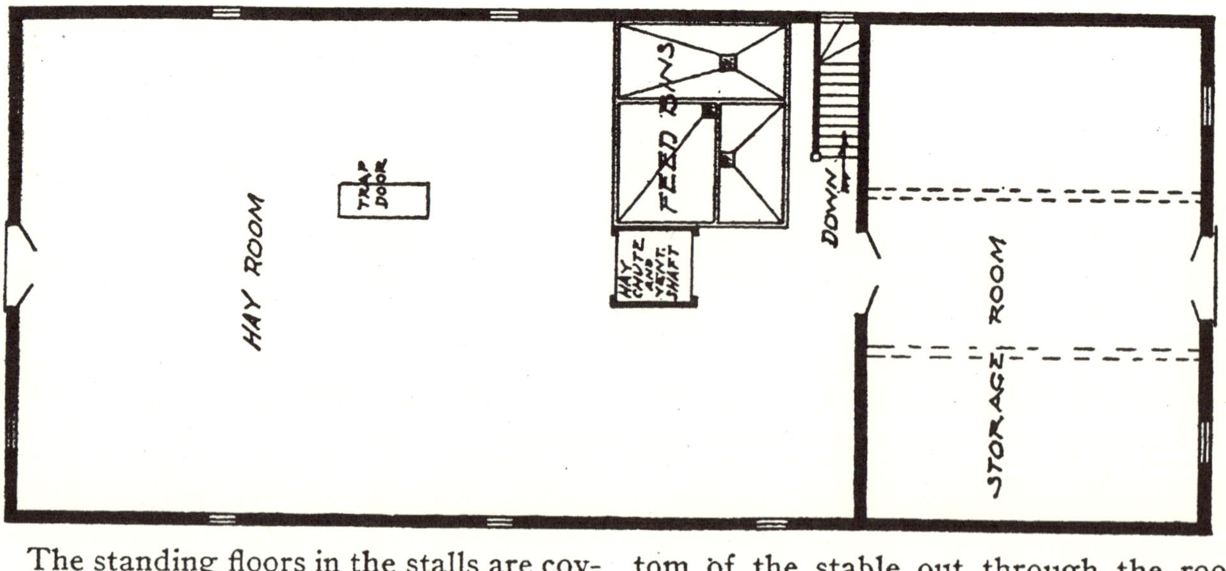

SECOND FLOOR

The standing floors in the stalls are covered with planks which of course are renewed occasionally, but the stall bottoms, partitions and cement mangers are there to stay. Horses do not gnaw such partitions and mice and rats are not troublesome, because they find no harbor of protection.

The balance of the barn is of frame construction with a curb roof. Another interesting feature of this barn is the two large galvanized iron ventilators with glass tops which conducts the foul air from the bottom of the stable out through the roof. The cross section shows the cement floor with piers and footings in profile and the cement work is indicated on the different plans and elevations.

Department of Horse Barns

HORSE BARNS

FARM buildings serve their purpose best when especially adapted to the specific use required of them. Horse barns should be different from any other building on the farm. The health and comfort of horses should be the first consideration, but convenience in attending to their wants and requirements, makes a close second.

All horse stables should be well ventilated. Every farmer knows that there is a great difference in stables in this respect. Some stables are so built that you would rather keep out of them if possible because they can't be kept clean. The smell of ammonia is always present and when the doors are shut it is very disagreeable. Imagine shutting a valuable horse up in such an atmosphere at night and expect to find him in good condition in the morning. Horses are the most expensive animals on the farm and the most susceptible to disease; hence, the first consideration in a stable should be to promote the health of the horses.

A horse stable should be cool and airy in summer and it should be warm and well ventilated in winter. The floor should be made in such a manner that it will not absorb the liquids from the manure, and there should be no crack to let these liquids down underneath to ferment and destroy the air in the building. Stable ceilings must necessarily be high enough to permit a horse to get his head up. Horses are warm animals, that is they contain body heat enough to warm a stable when conditions are as they should be.

Before starting to build put a little time on the study of ventilation. Read up on the circulation of warm air. Don't depend on others because they might not understand the particular conditions you are dealing with. It is well enough to ask advice, but get the information from different sources so that you may be able to sift the quality of your instructions sufficiently to keep the grain and discard the chaff. Don't blindly copy a stable that some one else has built without carefully considering whether it fits your requirements. A horse stable that works all right for one farmer is all wrong for another, because his horses may be larger, or has more of them, or he handles them differently. Some farmers have a lot of horses that they press into service in the summer time and turn them out in the yards and sheds to winter. Such farmers usually raise horses to sell and have more than they need at all times. Other farmers keep just what horses they need to do the work. They keep four horses or six horses the year round and they have no intention of altering their usual custom. But in either case a man can arrange a stable for a certain number of horses and build it accordingly.

In cold weather a stable big enough for six horses will not be warm enough if only two are stabled. If for any reason the stable is too large it is better to fill it up with cows in the winter for the reason that you cannot have ventilation without heat. On general principles it is more satisfactory to keep horses in a building by themselves and it is but little extra expense to do so.

When possible a horse stable should contain a carriage room that is reasonbly free from dust. Every man has or should have the ambition to keep a rig for the road that is decently clean. He owes it to himself and his family to provide a respectable turnout. A farmer's family depends for change and recreation on the opportunity to get away from home by means of the horses. They are judged to a very great extent by the apperance they make. You cannot get away from the fact that a person's social standing in the community is largely arranged for them by the opinion of others. No man is independent enough to stand alone. A man's usefulness in the community depends largely upon the appearance that he and his family make on dress occasions, and the appearance in turn depends very much on the horses, harness and wagons that they use when driving on the public road.

PLAIN HORSE BARN—A161

Cost of Blue Prints, $15.00

A plain straight-away horse barn with ten single stalls, five box stalls, feed room, harness room and vehicle room with a wash platform in the center is given in this plan. There is a driveway through the center wide enough to admit a load of hay or a load of straw, if so desired, but there are doors opening outside in the gable to pitch in hay and straw, either by hand or horse fork, so it would not be necessary ordinarily to drive inside with a bulky load, but a good passageway between horse stalls is a great convenience anyway.

PERSPECTIVE VIEW

This barn will easily accommodate fifteen horses and it will hold feed enough to supply them for a long time. The building is thirty-seven feet wide by sixty-eight feet long. It is set on a stone foundation with two rows of stone piers supporting the floor joists and posts which run to purlin plates.

There is a large vent shaft running from the stable ceiling to and through the hay mow with doors for throwing down hay or fodder as well as for ventilation. Grain in sacks can be hoisted up this ventilator shaft and conveniently dumped into feed bins which have hopper bottoms and spouts leading to the mixing room below.

In the driveway at one side of the mixing room door is a water supply pipe and watering trough with a hose connection to supply water to the wash room on the floor of the vehicle room.

The stalls are floored with a double

FRONT ELEVATION

thickness of oak flooring one and three-quarter inches thick slightly sloping to cast iron gutters, which run the entire length of the stall room on each side of the driveway. The first thickness of these stall floors is laid in hot tar, then two thicknesses of tar roofing felt is put on being well mopped over with tar, and this covered with the upper thickness of oak one and three-quarter inch flooring.

Where a great many horses are to be fed overhead feed bins are a great convenience. The bottoms may be made hopper shape as shown in the plan, or they may be level. A hopper, of course, is best, but with a flat bottom a little accumulation of grain around the edges at the bottom is all that remains when the grain stops running down the spout, and flat bottom bins are cheaper.

The main entrance doors are both wide and high. Unless the door is large enough it is sometimes difficult to get out. The door must have a good height because you want room for a carriage or a top buggy. We all have had experience in catching a buggy top on the lintel of a low door way. It seems to be the proper occasion for saying things. No builder likes to have such remarks made about him.

There is a good row of box stalls. It is difficult to plan a decent sized box stall in a small stable. They run into room too fast. Nothing looks so comfortable for a good horse as a roomy box stall. If the horses had their way about it there would be more box stalls, but it really requires about three times as much room to stable horse in this way. No man begrudges the room, but most men don't like to put up money enough to enclose it properly.

The ideal arrangement for stabling a horse is a big box stall with a good sized window for light and a door cut in half so that the upper part may be left open during the daytime to let the horse look out. A box stall shut up tight is a prison for a horse, they like to see things as well as other folks.

Some box stalls are fitted with rubbing boards. They consist of planks about two

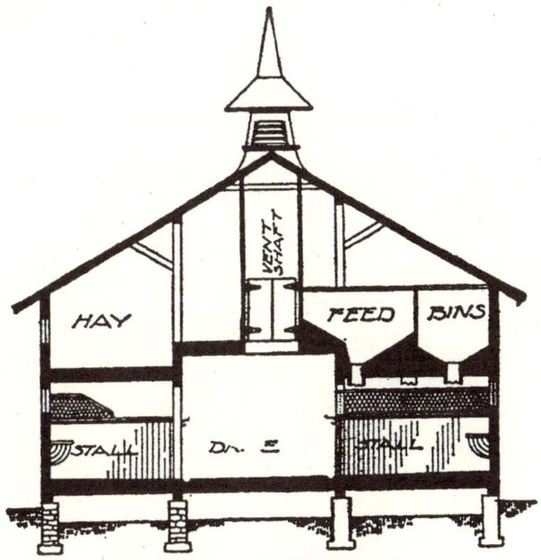

CROSS SECTION

inches thick turned edgewise to the horse and fastened to the sides of the stall just low enough down so the horse can't rub his tail. A box stall needs no floor and there should be no feed rack or manger. A box on the ground to feed oats is all the manger necessary. The hay should be put in at frequent intervals in small quantities placed lightly on the floor or bedding against the side of the stall. This way of feeding has often cured horses of chronic indigestion.

In building a stable it is a great deal better to find out all these little details and build accordingly. There are several reasons why box stalls are better than standing stalls with mangers. A horse loves his freedom. To understand this it is only necessary to watch a horse when you take the bridle or halter off.

One great defect in horse stalls as you ordinarily see them is lack of ventilation. It is quite common to see the inner walls of a stable in winter white with frost. The frost wouldn't be there if the stable was dry as it should be. It is not necessary to put in an elaborate system of ventilating pipes in a small stable. The windows and doors are sufficient if they are managed right. The breath of one or two horses is easily taken care of, but even in small stables such things often are neglected.

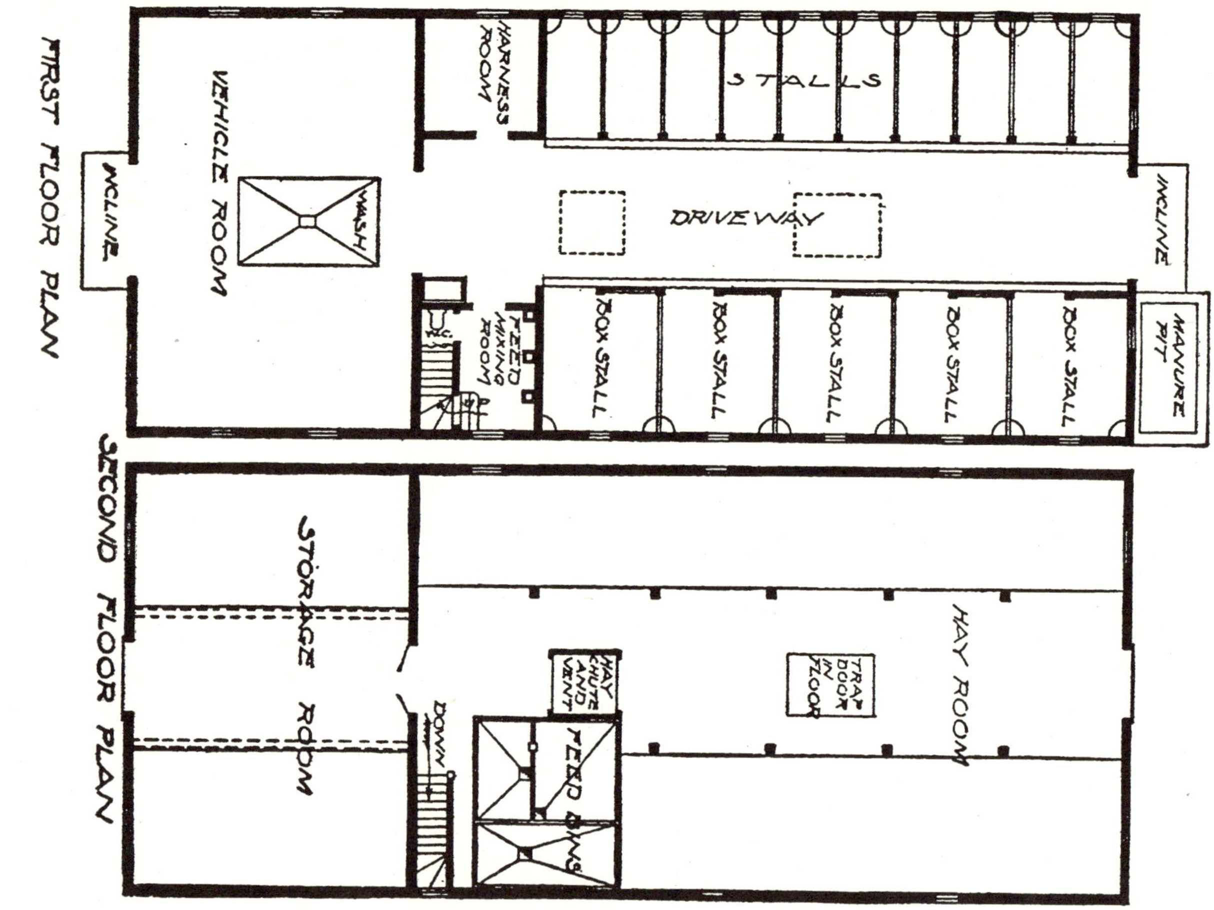

In this plan the carriage room is closed off from the stable which is right. The odor from the stable is a damage to the carriages and to the rugs. The stable should be warmer than the carriage room so the door works right from both sides.

CONVENIENT HORSE BARN—A133

Cost of Blue Prints, $15.00

Men who keep good horses will appreciate this plan. The arrangement of the stalls is convenient and there is a good carriage room in which to keep vehicles away from the stable part and out of the dust. Every farmer who takes pride in his horses likes to have a nice rig to drive, and it is impossible to have it without conveniences for keeping it clean. With a good carriage room and a good harness room there is no excuse for dirty buggies or an unsightly harness.

A feature of this barn that should attract special attention is the tool room. It is nine by ten feet in a front corner of the building with two good windows for light. There is a general work bench with a vise on one end and there are boxes to hold tools and supplies on the dark side of the room. The granary will be large enough or not according to the other buildings on the farm. Where there is a large grain barn for threshing a smaller granary in the horse barn seems to answer every purpose. The granary in this plan is placed right because it may be shut off

SIDE ELEVATION

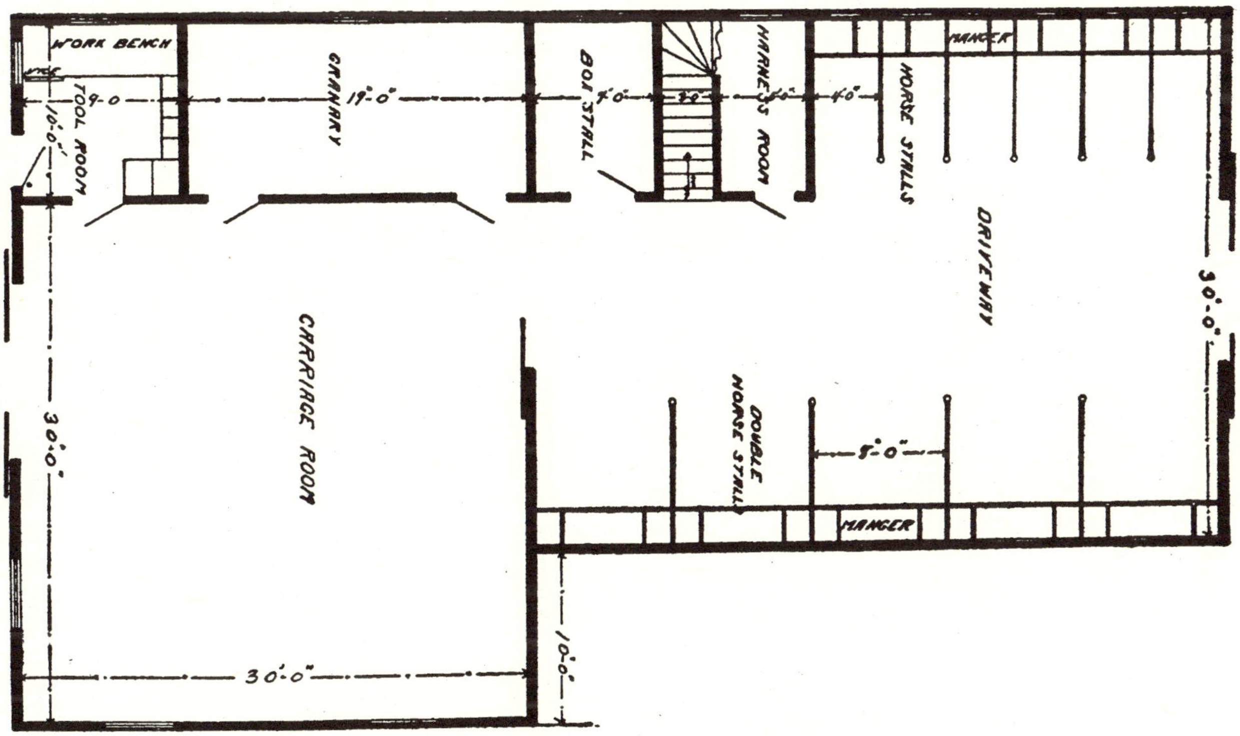

FLOOR PLAN OF HORSE STABLE

with two doors from the stable part, still

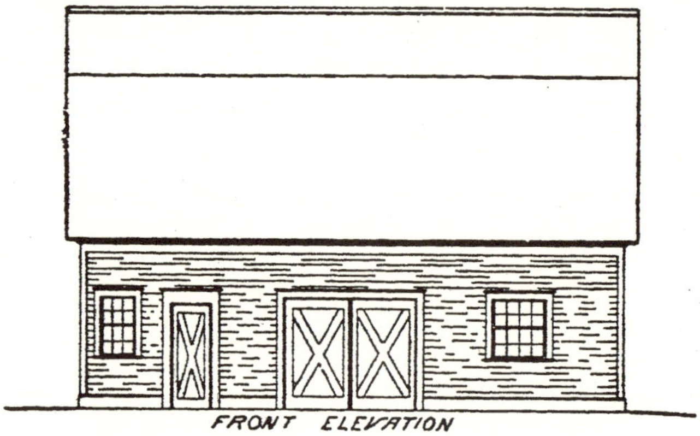

FRONT ELEVATION

it is not so far away as to make feeding inconvenient.

There is room overhead for a good deal of hay and straw. The hay carrier will bring this stuff from the back end pretty well through to the front.

It would probably be advisable to put a cement floor in this building.

There are a great many different kinds of floors put in horse stables, in fact one style or manner of building a stable floor seems to prevail in one part of the country when another county perhaps just a few miles distant seems to favor a very different way of doing things. In some neighborhoods you find nothing but plank floors, in other places it is all cement, then again you get into a neighborhood where there are no floors at all.

EIGHT HORSE STABLE—A124

Cost of Blue Prints, $5.00

A small cheap horse stable is shown in plan (A124). It sometimes happens that

SIDE ELEVATION

a separate stable for horses is necessary because of the manner in which the other

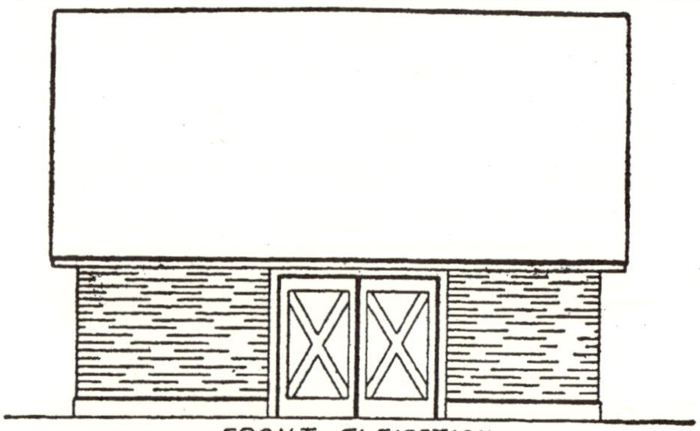

FRONT ELEVATION

buildings are constructed and occupied. This little stable will accommodate eight horses and there is room enough overhead to hold the straw for bedding, but it would be necessary to provide the feed from

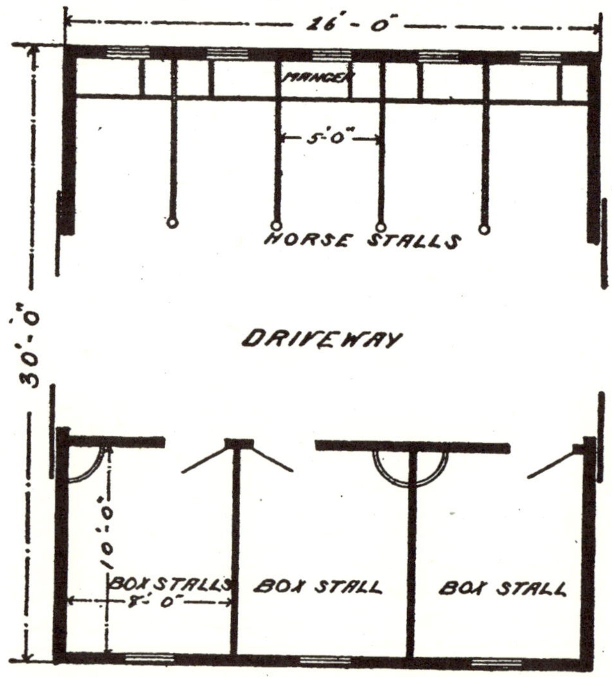

some near-by storage. It is not necessary to put a floor in this stable unless it be on the side where the open stalls are built. But a good many horse stalls have stiff clay pounded in and there are plenty of horsemen who prefer such stable bottoms.

CITY STABLE FOR TWO HORSES—A114

Cost of Blue Prints, $5.00

A very neat carriage house is shown in plan (A114). It is intended to house two horses and have room enough for a couple of carriages. The building is supported by a stone wall three feet in the ground

FRONT ELEVATION

and one foot above the ground to keep the floor well up, but the height of course must depend on the nature of the ground and location in reference to the street and driveway. It is not desirable to approach

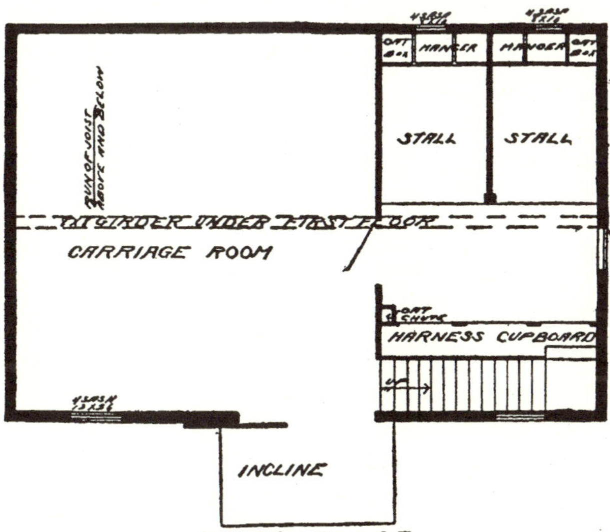

FIRST FLOOR

the main doorway by a very steep bridge because it is often necessary to run carriages out and in by hand. Of course if it is necessary to set the floor up the driveway may be raised accordingly, this however very often runs into considerable expense.

The way a driveway approaches the stable affects the appearance of the stable a good deal. Generally a pleasing effect may be obtained by a curved driveway where it is kept neatly trimmed at the

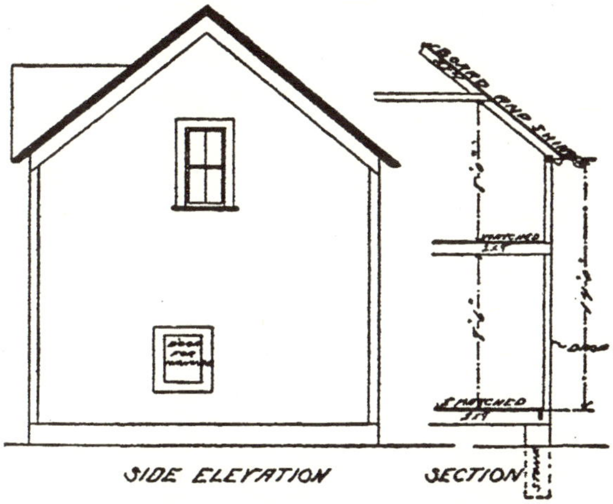

SIDE ELEVATION SECTION

sides. If the driveway is gently rounded and the edges kept about two inches lower than the sod it is easy to maintain a clean track and a well defined edge without putting a whole lot of unnecessary work on it. The lawn mower will trim

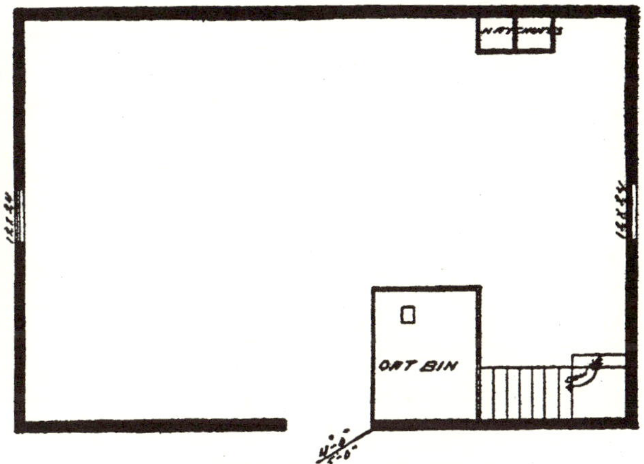

SECOND FLOOR

the grass and a spade used once a month will keep the edge of the drive in good shape.

The floor of this carriage house is made solid by running a heavy girder lengthwise of the building through the center. Joists are carried from the sills to meet the girder. The floor is double, the first layer being an inch thick dressed on one side to make the boards even in thickness, is laid diagonally. On top of this is laid a layer of felt roofing mopped with tar, both unneath and on top. The upper floor is one and three-eighths matched hard pine.

In the stalls two inch planks are laid lengthwise, having an incline of two inches in the length of the stall. These planks are nailed to one cross piece in the middle and another cross piece a little thicker under the manger, but the nailing is not very solid because stable planks soon wear through and it is necessary to turn them end for end, sometimes within a year.

SMALL BARN WITH CEMENT FLOOR—A112

Cost of Blue Prints, $5.00

This barn is twenty-two feet wide by thirty-four feet long and has a cement floor cushioned with cinders the whole

West End.

size of the building, but the standing stalls have a plank floor running lengthwise of the stall over the cement. These planks

South Side

are not fastened except to two cross pieces —one under the manger is a two by four laid under the plank to give them the proper pitch. Another cross piece an inch thick is placed in the middle to strengthen the plank, back of this the planks have free ends which facilitate drainage back to the gutter and makes it easy to remove the

East End.

floor if the planks should split or wear out.

The box stall may have an earth floor, if so desired, three or four inches thick, made of good stiff clay wet down and

North Side

tamped level over the cement. Some horsemen prefer a cement bottom with a foot or two of straw; either way is good enough if the horses have the right kind of care.

The oat bin is in the hay loft and the

corn bin may be put there too if the space on the carriage room floor is needed. By

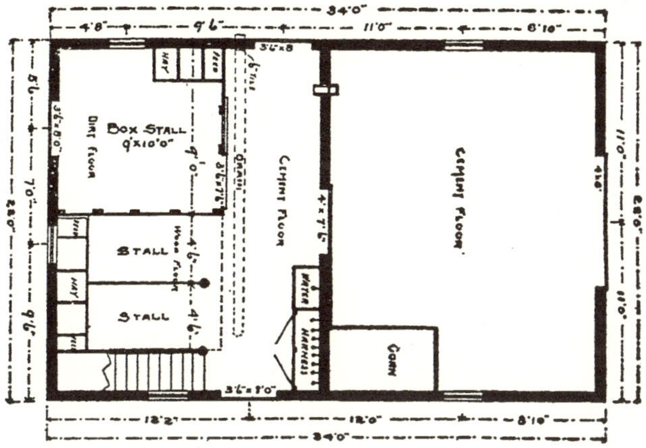

having the feed overhead and chutes for the different kinds of feed to the floor below, feeding is made easy.

Sliding doors usually are preferred for a horse barn, and half door for ventila-

Second Floor

tion is a good thing. A horse will stand for hours with his head out of such a door with evident satisfaction.

VILLAGE STABLE WITH CELLAR—A116

Cost of Blue Prints, $5.00

A very neat, attractive stable for a city or village is here given. A good stone wall is laid down below frost, or it may be carried a little deeper and the part under the

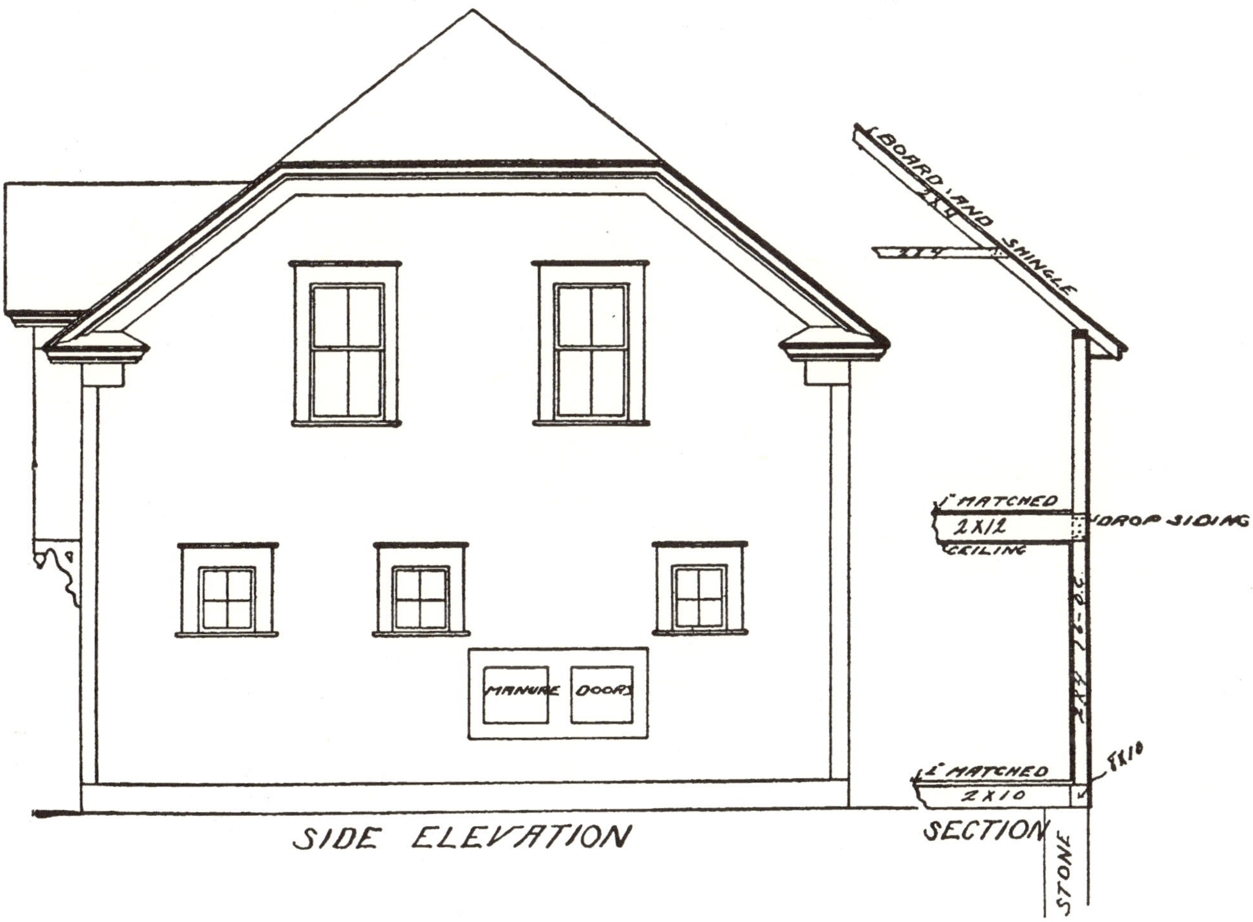

FRONT ELEVATION

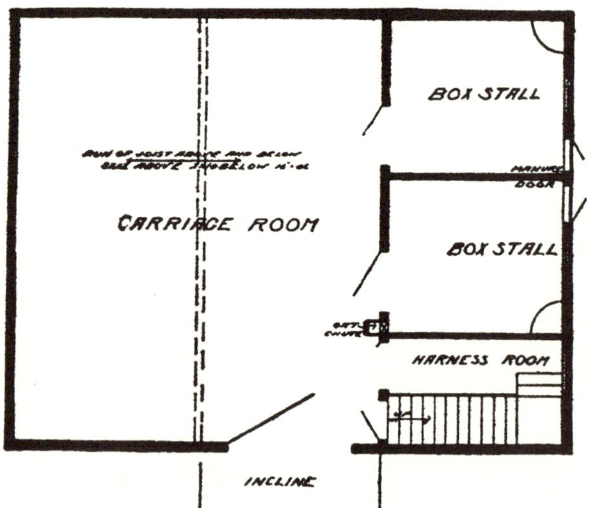

FIRST FLOOR PLAN

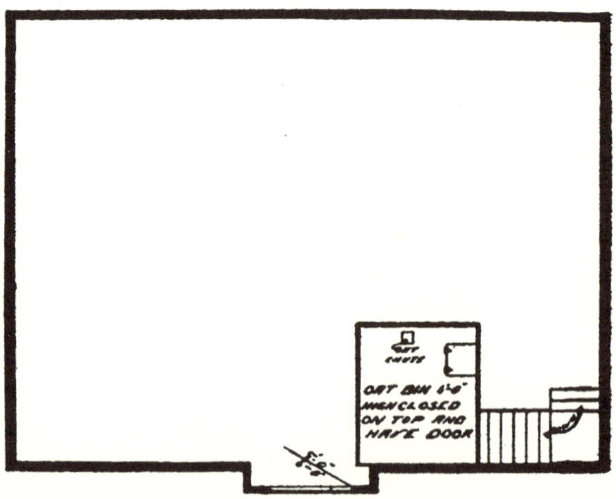

SECOND FLOOR PLAN

carriage room excavated for a cellar, but in this case the retaining wall would be necessary on the stable side because the box stalls are supposed to have an earth floor. Any way, you don't want horses over a cellar. The elevation is pleasing because it is not exactly plain, still there is no great additional expense in building a roof like this or in the little projection from the upper door in front. There is storage room above for hay, straw and oats, and the upper door is wide enough and high enough to admit the supply easily. The doors to the box stalls should be made in halves so that the upper half may be opened and the lower half closed.

NEAT BARN FOR HORSES—A156

Cost of Blue Prints, $5.00

This plan is a very neat arrangement for a city or village lot where two horses are kept together with the necessary carriages and harness equipment. The carriage room with rack for washing buggies is about as well arranged as it could be and the harness room being under the stairway occupies as little space as possible. Another nice arrangement about this barn is the location of the manure door. The stalls may be cleaned and the manure thrown out at the back as far away as possible from the carriage entrance and from the side entrance to the man's room.

SECTION END ELEVATION

SIDE ELEVATION

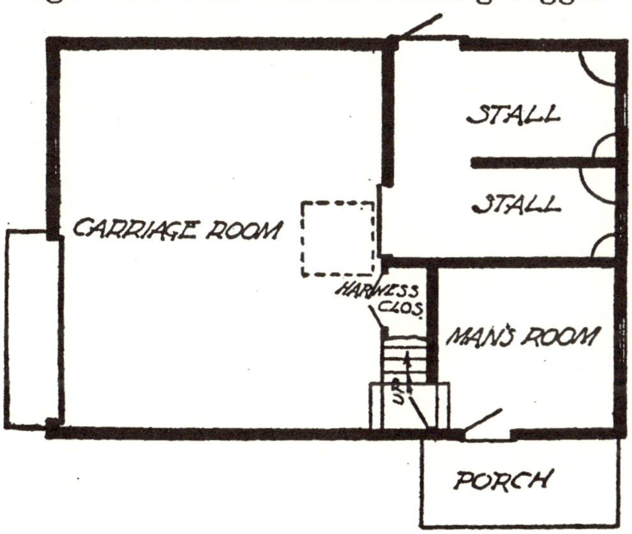

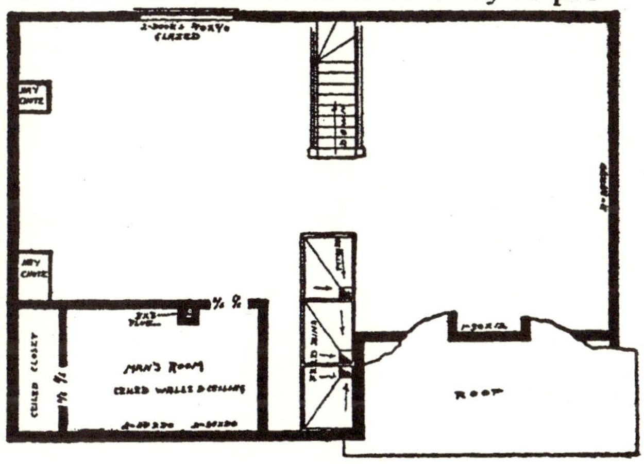

SECOND FLOOR PLAN

Where horses are kept in town there is usually a lawn that requires attention and more or less garden work to look after. For this reason it is often necessary to keep a man and it is desirable to have a room that he can occupy outside of the house. It is a good thing to have a man sleeping in the stable where valuable horses are kept so this arrangement works first rate for several reasons. It is hard to keep good men even in town unless they have comfortable accommodations. This building is thoroughly well constructed and the room arranged for the man is more comfortable than similar rooms in some houses.

THREE STORY HORSE BARN—A117

Cost of Blue Prints, $10.00

A bank barn for horses, thirty-two by thirty-six feet, is shown in this plan. The stable is in the basement and on the main floor there is a driveway with a corn crib on one side and bins for grain on the other. Above this main floor is mow room for hay and straw.

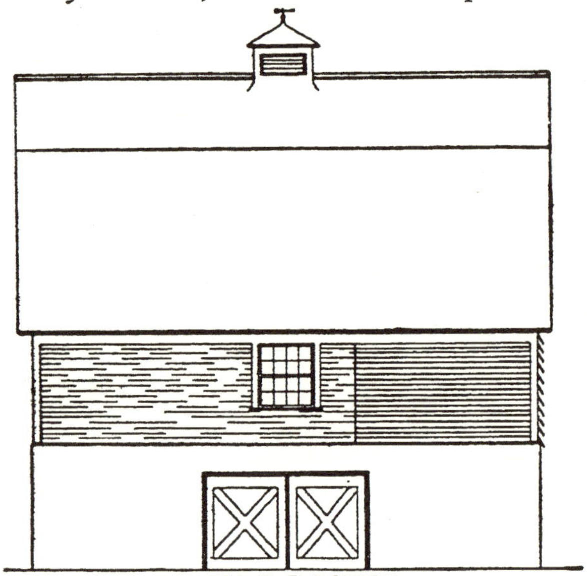

FRONT ELEVATION

SIDE ELEVATION

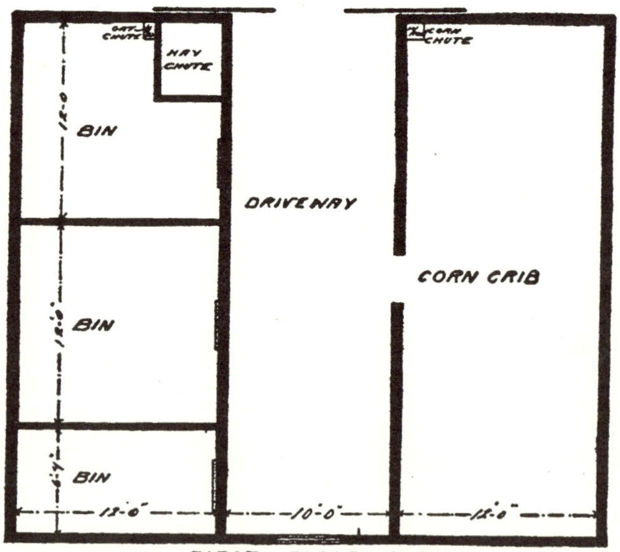

FIRST FLOOR PLAN

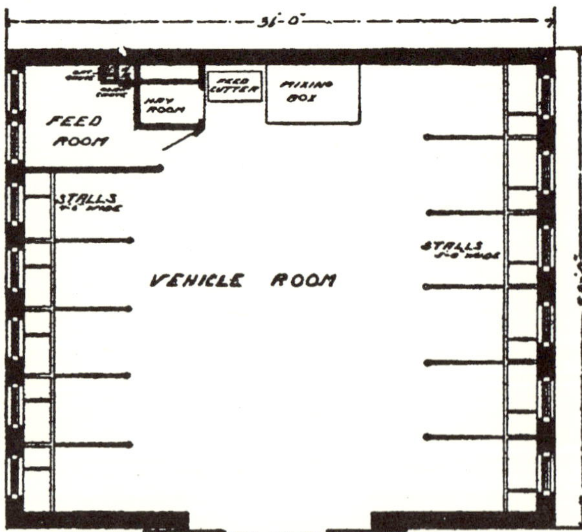

FLOOR PLAN OF HORSE STABLE

This barn will furnish stabling for eleven horses in the basement, besides a feed room which is connected with the grain bins above and the corn crib as well as hay mow by means of chutes. The hay and oat chutes are perpendicular and pass straight down from the loft and from the grain bin to the feed room below. But the corn chute is built diagonally across under the main driveway floor to carry it over to the feed room. The reason for this is that all feed rooms should be shut off from the stable with a good door. Most of us have had experience with horses getting loose at night and eating more grain than was good for them. This corn chute is twelve by fourteen inches, which is small enough, considering that it is a slanting chute. All grain chutes when built in this way require to be larger than when placed vertical, because there is more friction in the passage of the grain in coming down.

The corn crib is ventilated on three sides by using narrow strips nailed to cleats slanting outward. This will answer for corn that is reasonably dry, but unless the weather is favorable it is not a good plan to fill a bin like this full of corn without some kind of a ventilator in the middle.

CHEAP HORSE BARN—A113

Cost of Blue Prints, $10.00

A small barn with two double stalls and one single stall with standing room for another horse is offered in this plan. The barn is twenty-six feet wide and thirty-two feet long, one half of which is partitioned off for a stable and the other half

WEST END.

EAST END.

SOUTH SIDE.

NORTH SIDE.

is kept for carriage room and storage. There is no foundation under this barn except stone or brick corners and center supports, but it is a good plan to put a board around under the sill and bury the lower edge in the ground.

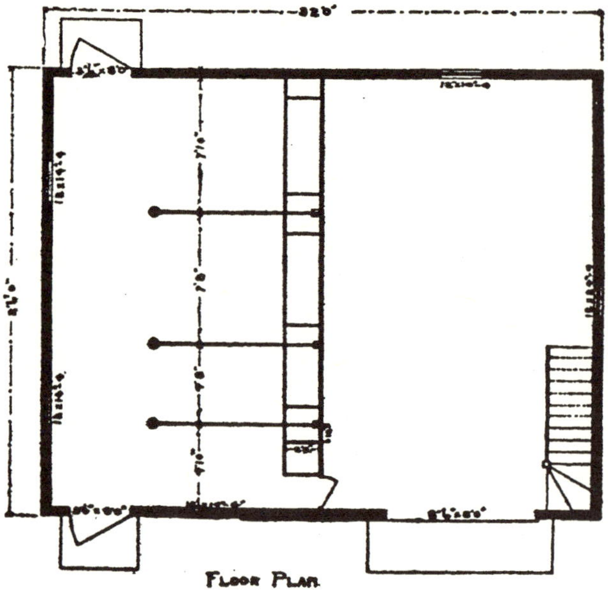

Floor Plan

A barn that is open underneath makes a harbor for rats. It is better to have it boarded up. The stable doors in this plan, both at the north side and at the south side, are cut in two so the upper half may be opened for air and ventilation and the lower one remain shut to keep the animals from going out and in. The plan is as simple as possible to make a barn and still have it look well. It is large enough to be of some use and it has quite a loft for hay. A cheap little barn like this sometimes answers the purpose as well as a more expensive one.

It is a barn that would suit the average merchant who is engaged in other business besides actual store keeping. There are many such men who have a pair or more of horses for teaming purposes and who want more stall room than the ordinary small horse stable provides.

ANOTHER CHEAP STABLE—A132

Cost of Blue Prints, $5.00

Plan A132 is a small carriage house which may be built at very little expense. It often happens that a man wants to keep a horse for his own driving when he don't care to put a great deal of expense on the stable. It is a mistake in such cases to build a cheap looking affair because a man is never satisfied with it and it injures a person's property. It is just as easy to build an attractive stable, one that is well proportioned and well designed because if rightly laid out it costs but little more than a poor looking affair that has a cheap appearance. It is all right to build cheap

Front Elevation

Side Elevation

if nobody finds it out, but we often see miserable structures that give away the owner's ambition.

Here is a stable that costs very little to build but you never would know it, especially if it is neatly painted and nicely kept both inside and outside as it should be. There is sometimes more genuine satisfaction in a cheap building well cared for than in an expensive structure that is permitted to go to seed.

The size of this barn is eighteen by twenty-four feet. Its attractive appearance is due more to the shape of the roof than to the general design or to any other one feature. All village barns should be placed carefully on the lot to look well and so they will not annoy the neighbors.

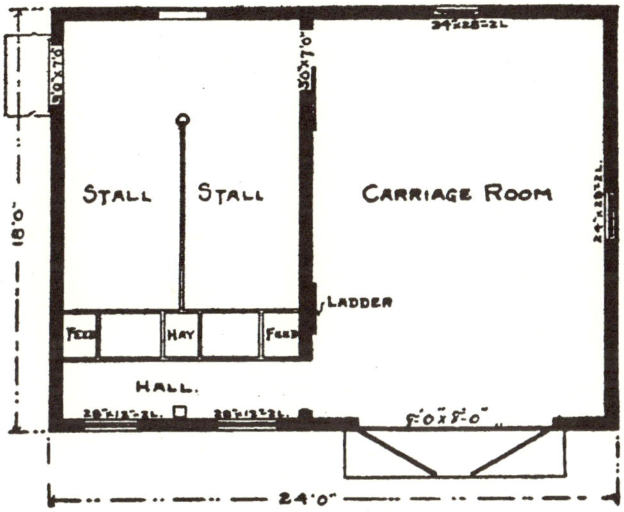

FLOOR PLAN

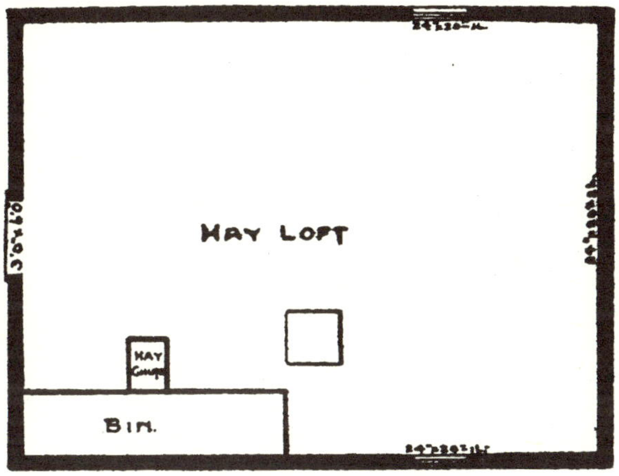

SECOND FLOOR

SMALL BARN FOR A VILLAGE LOT—A111

Cost of Blue Prints, $3.50

This is just a little affair, only eighteen by twenty feet, but it is big enough to hold four horses and leave room for a wagon on the storage floor. There is also loft enough to mow away three or four tons of hay. It is not necessary to make a very deep foundation for a little barn like this. If the ground is leveled and three or four courses of brick laid around under the sills the building will set all right probably for a good many years. Many small barns are just blocked up on

WEST SIDE.

SOUTH END.

stones placed at the corners and one or two places along the sides but this is objectionable because it makes a harbor underneath for vermin. The foundation should have some air but air enough will penetrate through the chinks between the bricks if they are laid without mortar.

The construction of this little barn is about as plain and simple as it could be and still have it look right when finished. Nobody likes a cheap looking building, but no one objects to a goodlooking building if they get it cheap. The problem is how to build what will be satisfactory in a few years' time. Sometimes an inexpensive building may be shaded with trees or screened by vines in such a way as to give it a presentable appearance even in winter. An evergreen or two planted along the side, if there is plenty of room, makes a great winter addition to the looks of a stable. Grape vines usually do well if suspended by wires from the eaves, but grape vines should never be tacked close to the side of a building, they need air on all sides.

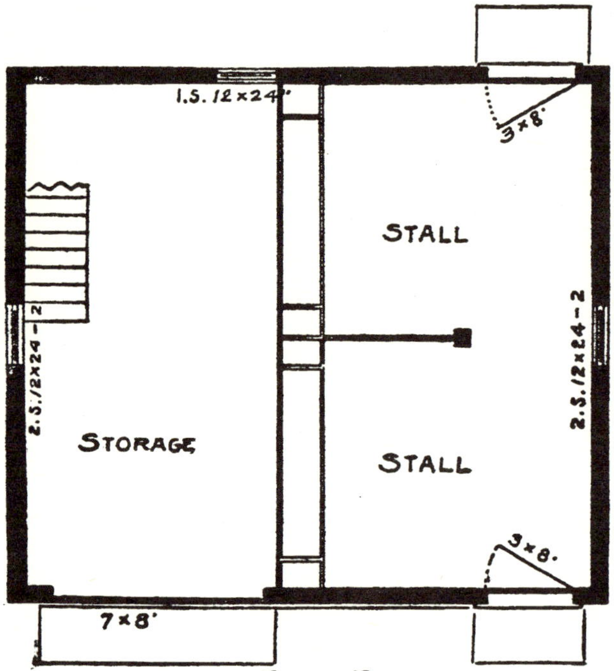

FIRST FLOOR

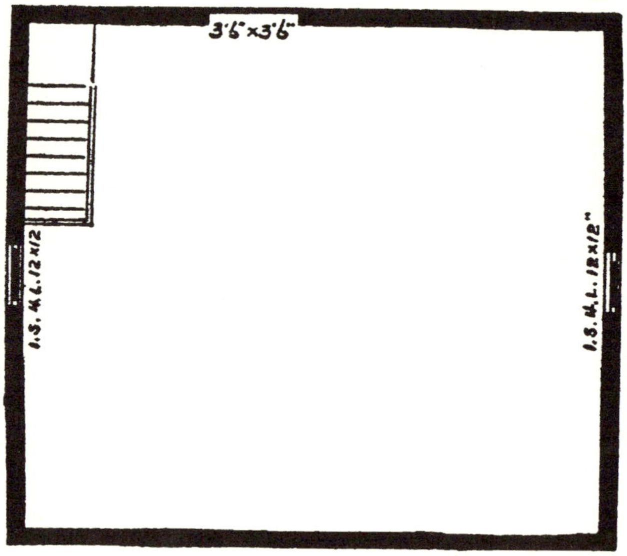

SECOND FLOOR

SERVICABLE BARN—A172

Cost of Blue Prints, $10.00

This is a small barn, twenty feet by thirty-two feet, and contains a carriage room thirteen feet by nineteen feet, which has large double doors in front that will admit the largest size carriage, a wide single door to the horse stable, and a stairway leading to the upper floor, which is for the storage of hay, feed, etc., and a man's room if it is desired.

This barn contains two single stalls and a box stall. Each stall has a direct window, which is high enough from the floor to avoid too much draft on the horses and it protected by a wire mesh guard.

This barn has been designed for utility and is practical in every way. The arrangement is convenient, and it is of a neat appearance on the outside. If painted a stone grey, with all trimmings and cornice work painted pure white, it would be a credit to any neighborhood.

The carriage room has a cement floor, which is slightly pitched from all directions down to the center, where it is provided with a floor drain. This will admit the carriages to be washed any place in the room without injury to the floor or the side walls which are wainscoted with Portland cement to a height of two feet six inches.

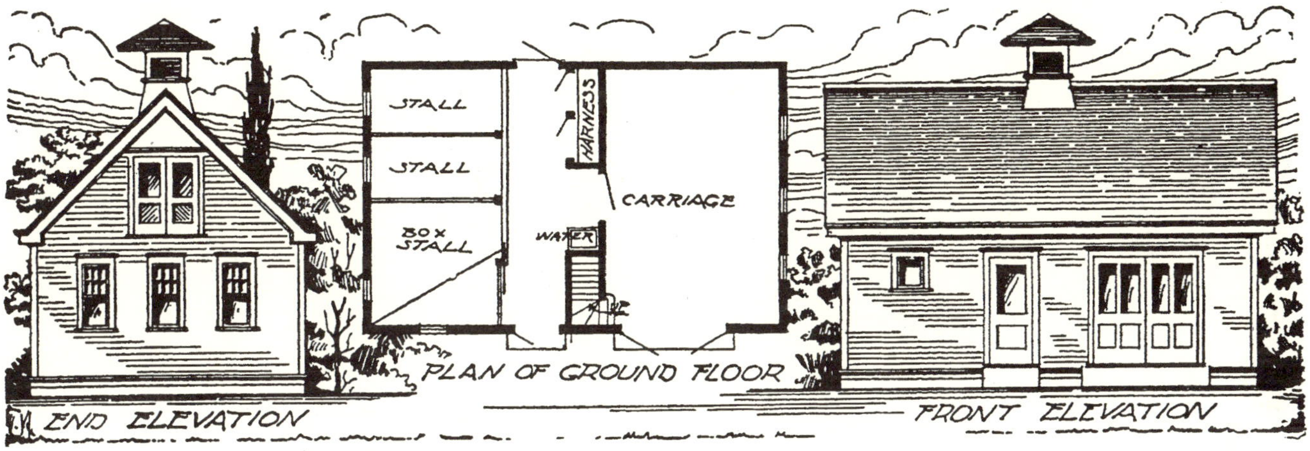

All the walls of the first story and ceiling are finished with clear southern yellow pine, beaded ceiling, with two coats of hard oil. This makes a very pretty effect for a stable and it is at the same time very serviceable. The stall floors are of double thickness one and three-quarter inch floors. The first floor is tongued and grooved, tightly laid, and then covered with hot tar. The upper floor is then laid and has slightly beveled edges, so that when laid the boards will fit tightly together at the bottom and leaving about an eighth of an inch crack on the top surface, which is then filled with hot tar. This construction makes a very durable and sanitary floor. The entire stall floor is pitched slightly to the rear to a cast iron gutter with perforated cover and connected with the catch basin and sewer. The second floor has ample storage room for a winter's supply of hay and feed for three horses and is of strong construction. The roof is of shingles and the ventilator gives the building a complete appearance.

SEPARATE HORSE BARN—A129

Cost of Blue Prints, $5.00

A small, convenient horse barn, twenty-one by thirty-two feet in size, with considerable mow room is shown in this plan. Such a barn is very convenient on some farms where for good reasons it is found best to keep horses in a building by themselves.

There are a good many farmers who object to stabling horses in the same building with other animals, because they don't seem to mix just right. Horses are different in their habits from many other domestic animals and it seems right and proper to give them a building to themselves when possible. Besides, it often is more convenient to have a small horse barn near the house to save steps in doing the chores. A horse barn is in use every day in the year, while on many farms the cattle barns are not used much in summer. Then again a horse barn properly cared for has no disagreeable odor. It may be near the house without causing annoyance. Very often women have driving horses of their own and they like to look after them themselves to a certain extent, and they very much prefer to have them within easy reach. Also in case of fire there is a further advantage in having farm buildings separate.

The old English plan was to scatter farm buildings far enough apart to pre-

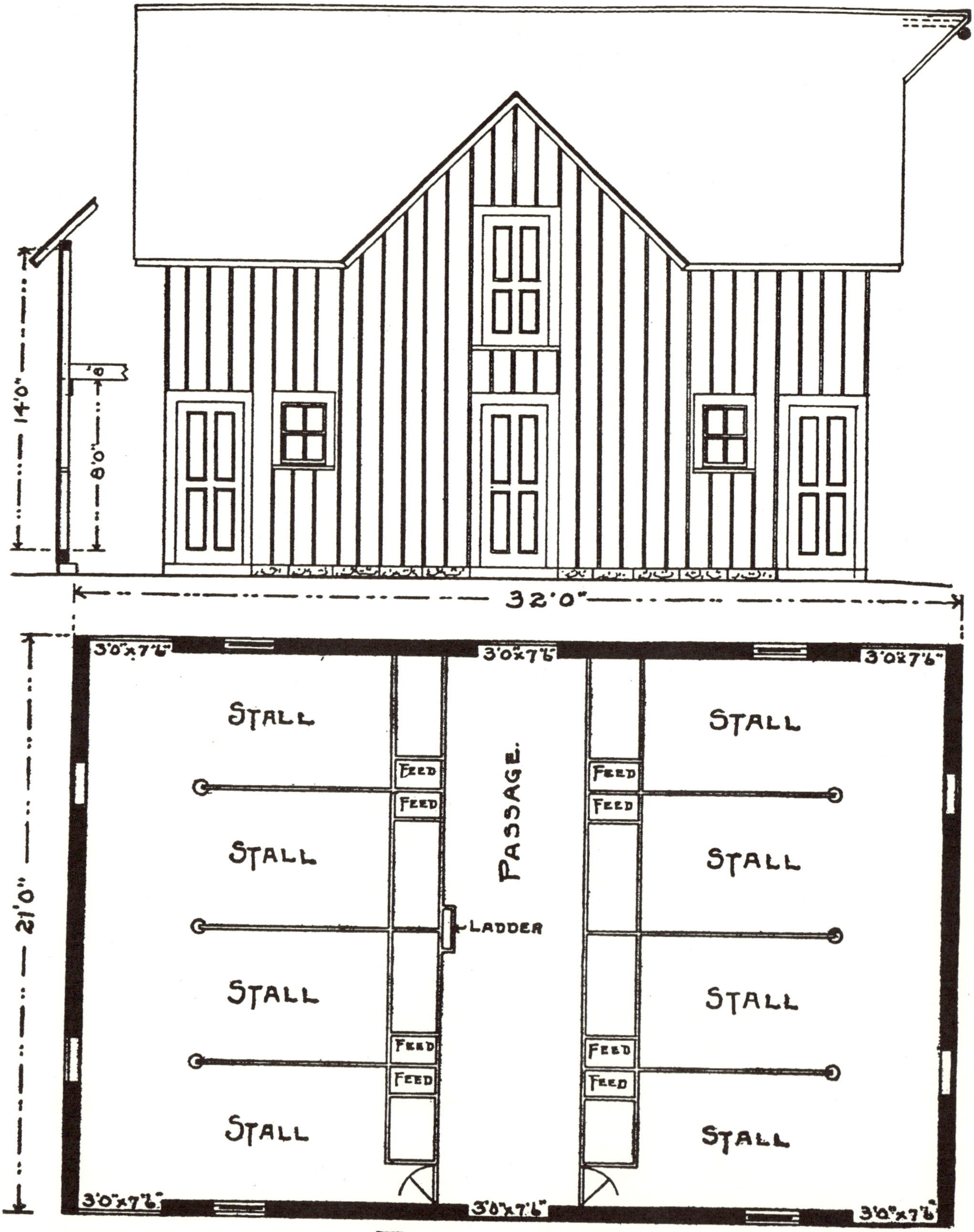
14'0"
8'0"
32'0"
21'0"
3'0"x7'6"
3'0"x7'6"
3'0"x7'6"
STALL
STALL
FEED
FEED
PASSAGE
FEED
FEED
STALL
STALL
LADDER
STALL
STALL
FEED
FEED
FEED
FEED
STALL
STALL
3'0"x7'6"
3'0"x7'6"
3'0"x7'6"

FLOOR PLAN.

vent a general conflagration in case one should take fire, but farm labor is less expensive in England and everybody knows that it costs more to care for animals housed in separate buildings because of the running back and forth and because you haven't a great big storage room all under one roof where feed may be hoisted by horse power and returned to the feeding floor by gravity.

In building a little horse barn like this it is better to put down good foundation walls reaching below frost. By making the passageway floor about three inches higher than the floor behind the horses an incline will be provided sufficient to keep the feed room floor dry as well as to give the necessary drainage slope to the standing floor for the horses.

Most horsemen prefer to floor the horse stalls with planks, whether the bottom is cemented or not. This may be done before the partitions are put in, but it is better to plank each stall separately. In either case select planks two inches thick with tongue and groove matching and lay them with coal tar between. Give the floor a slight incline, say two inches fall in a distance of eight feet.

It is much better in a barn of this size, built for this purpose, to cover the whole bottom with a cement floor, cementing tight up against the walls all round and leaving a slight depression behind the horses; a sort of rounded open drain not more than an inch deep and slope the drain to the manure door so it is easy to wash the stalls and sweep the water out doors. A horse stable after this order will be found very convenient on any farm whether other buildings are calculated for horse stabling or not.

LITTLE VILLAGE STABLE—A135

Cost of Blue Prints, $5.00

The little barn, eighteen by twenty-four feet, as shown in the plans and elevations is a very satisfactory design and can be used in either village or city. It is not expensive, in fact, it is probably as cheap as any satisfactory structure could be. It is better not to take up room in such a small barn in building a stairway, as the upright ladder placed against one of the partitions answers the purpose very well. To keep the cold from blowing down through the opening a light door with a pulley, cord and counter weight may be made to shut over the opening.

FRONT ELEVATION

SIDE ELEVATION

If there is a boy in the family he will find a way to rig up a workbench in the front corner of the carriage room between the door and the first window. It is easy to encourage boys to work with tools, especially since the graded schools have taken up manual training. The

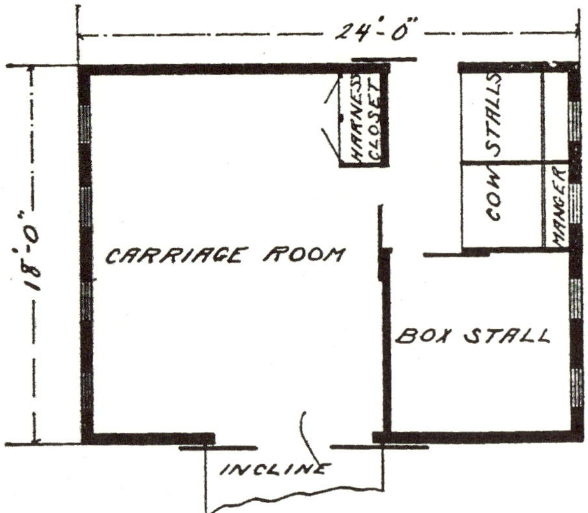

FIRST FLOOR PLAN

schools have added tone to the work, boys don't consider it labor now, it is part of their education and it is an important part, too. Truth may be taught in a more thorough manner through mechanics than by any other means. The principle of learning a thing by doing it is just as valuable now as it was in Froebel's time.

As a general thing a boy's work with tools is not very valuable when judged from a mechanical standpoint or from the amount of money that the finished product would bring, but it very often has a great educational value to the boy that is little appreciated by the older members of the family. The fundamental principles of mechanics permeate all nature. Animals are built on the best mechanical principles. There is a very close connection between mechanics and nature. Mechanics point the way to the connecting link between natural phenomena and commercial success. Mechanics and mathematics also are very closely related, but the natural live boy loves the one and hates the other. No woman wants a boy tinkering in the house, but he can spend many happy hours in the barn without disturbing anyone.

ENGLISH CARRIAGE HOUSE—A99

Cost of Blue Prints, $10.00

Small artistic stables are more common in England than they are in the United States, possbily because the country is older and the people have had more time to develop an artistic taste in such matters. An English gentleman likes to keep his cob and cart. He wants a good smart turn-out that presents a respectable if not a dashing appearance; then he likes to have things in keeping at home, so he maintains a very neat carriage house and stable.

Some ot these carriages houses are older than the proprietor but you would never know it to look at them. They are kept in such repair and they nestle amongst the hedges and trees in such a pretty homelike way that you never think about their age or intrinsic value. You get the impression at once that they are proper and proper goes a long way in England. You don't wonder that they have very neat stables just the right size and that they appear modestly retiring away to the back end of the pretty garden. It just seems to come natural. Their great, great grandfather or their double great uncle did the same thing long before they were born so all they have to do is to follow precedent.

The English carriage house of today was built after hundreds of years of experimenting until the location of every plank, the size and direction of every door and window was determined without any further question in regard to the possibility of the slightest improvement. It is put back on the lot in the furtherest corner from the house. The approach to it is through an arched or pillored opening in a beautifully well kept hedge. The driveway is not straight. English gardeners keep just as far away from straight lines as they possibly can. Somebody discovered in the time of King Alfred that curved paths and roadways in gardens were proper. Some of the old enthusiasts went a step too far and got them crooked. This was frowned on for a century or two until succeeding generations pulled some of the kinks out by injecting a few liberal doses of English conservatism so that now after a good many generations the driveway from the lane through the back of the lot to the stable is gently curved. The stable also is partially screened from

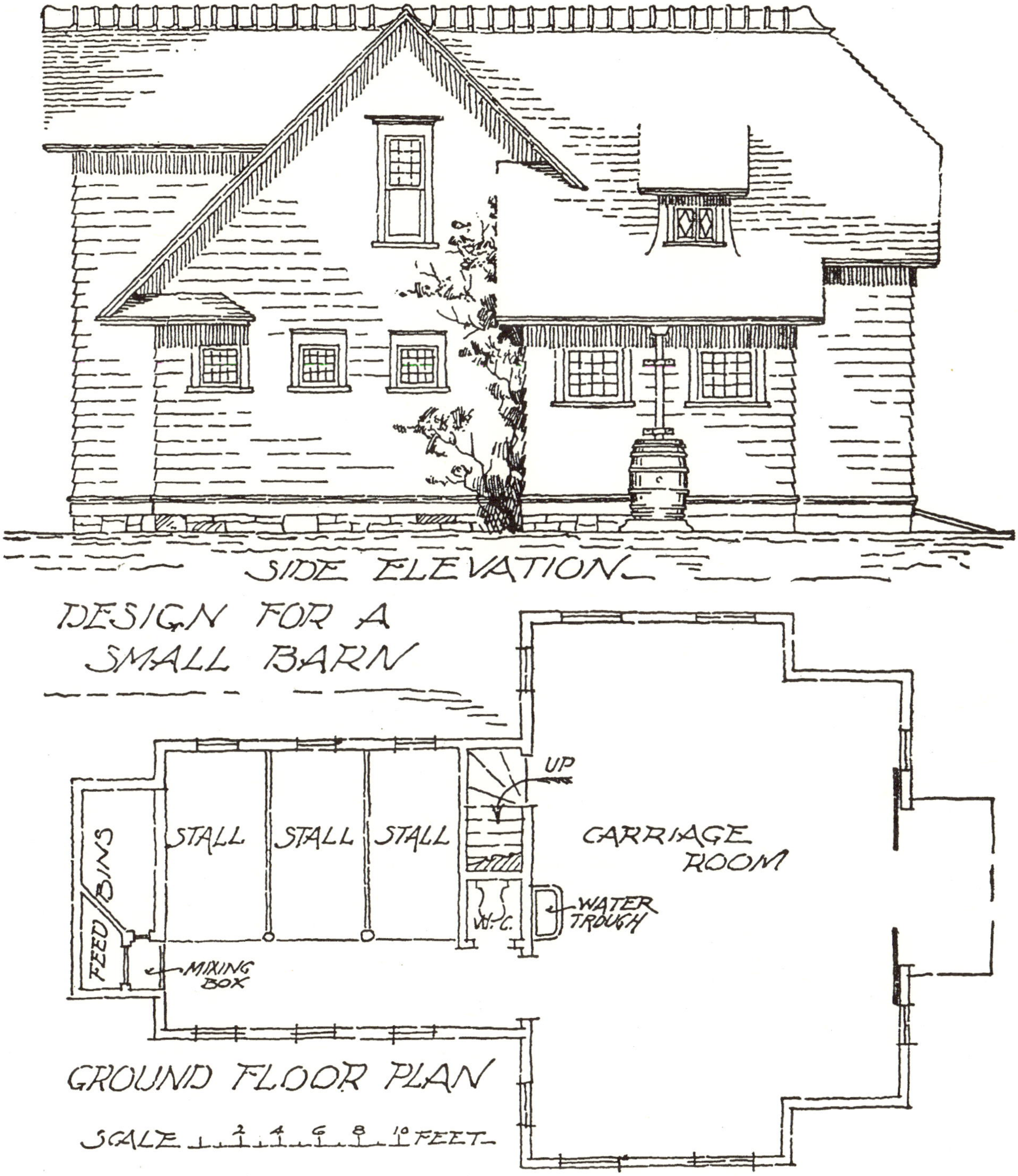

view by hedges, vines and trees: This is proper in England, it is good sense in any other country.

The difficulty of doing things just right in the United States is that we are in too much of a hurry to get satisfactory results. We get ready to build a stable one day and have the material on the ground before breakfast the next morning. We haven't decided where to put the thing so we go out with the carpenter harboring the idea that his time is going on and that while we detain him he is not engaged in sawing or hammering. For economy sake we must decide instantly. The street line is guessed at and the barn placed just a little inside. After it is up and the workmen have gone there is plenty of time to think it over and regret not having done some things differently, but the barn is up now, it has cost a little more than we counted on, they always do cost more than we expect, and we always expect they will when we start in, but at any rate we haven't any time or money now to change things or even level off the ground properly. We haven't figured on a curved driveway, that is all nonsense, but we lay down some planks to keep us out of the mud. The finish is not satisfactory to ourselves or anybody else, but we have a barn and we have secured it in characteristic American hustle fashion so we ought to be satisfied.

The plan (A99) shows the general arrangement. There is a room partitioned off in the gable upstairs for the man. A stairway going up from the carriage room lands in this upper room. The feed bins at the back of the stalls connect with the storage bin on the upper floor by means of spouts as indicated. There is a carriage room that is large enough to look well and to accommodate a number of vehicles. Instead of having a harness room there are pegs for harness in a corner of the carriage room and the harness is covered with curtains hung to a wire overhead.

SMALL SUBURBAN BARN—A215

Cost of Blue Prints, $10.00

Showing a barn erected for the accommodation of two horses and two cows. It is 26 feet wide by 34 feet long and is constructed of frame with cement plaster "rough cast" exterior wall and stained

shingle roof. It is set on a foundation of concrete which runs one foot higher than the ground floor, thus avoiding all dampness from the floor and ground coming in contact with the wood construction, which would otherwise soon decay.

The sill of a barn is always the first member to start decay on account of dampness absorbed from the ground, and if this essential member of the structure is rotted away the balance will soon follow. Any method employed in the con-

NORTH ELEVATION

struction of a building that will lengthen the life of the sills is worth looking into.

The roof of a building is justly considered the most important part; a building can be built without much of a foundation by setting it on posts or masonry piers, and for a few years perform all its require-

WEST ELEVATION

ments, but the roof can never be omitted; and like the sill, it must be kept in good

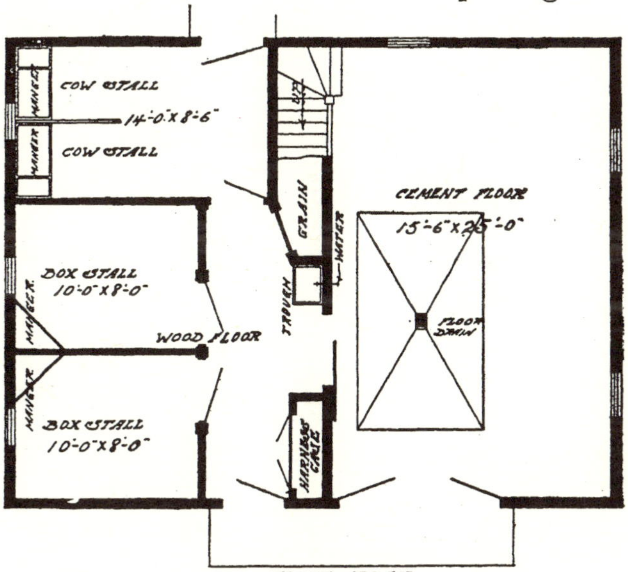

FLOOR PLAN

condition or the building will soon go to ruin. One of the most important parts of the roof is the gutter or cornice which, if made without projection beyond the vertical walls, is of little value, as it will al-

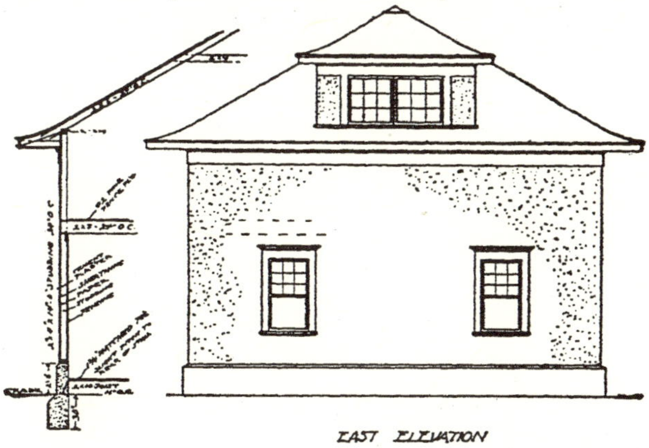

EAST ELEVATION

low the rain to run down the sides of the building and soon make the walls look weather worn and streaky from the dust which is washed down the walls with the rain. A well built cornice with a good projection not only avoids this trouble, but also protects the walls from the hot summer sun as well as giving architectural grace to the design.

The outside walls of this building are constructed of 2 inch by 4 inch studding, 16 inches on centers, sheathed on the outside with matched sheathing, then covered with waterproof building paper; then 1 inch by 2 inch furring strips placed 16 inches on centers (or over each studding); then lathed and cement plastered. This not only makes a very durable wall, but is warm in winter and cool in summer.

The carriage room and cow stalls have cement floors and the box stalls are of plank. There is a large hay room on the second floor, a grain bin under the stairway, and a harness case near the horse stalls.

GOTHIC BARN—A181

Cost of Blue Prints, $10.00

If the horse barn is near the house and if the house has a steep roof the barn should have a similar roof to be in keeping. We often see a house of one style and the other buildings nearby built on entirely different lines. If the house is new and the other buildings old there is some excuse for such incongruity, but in most cases the house is built first and the barn is added to the lot some years afterwards. In the meantime some architectural fad has taken possession of the neighborhood and every building erected must bear the marks of the new fashion.

There is too little originality in building. It is much easier to follow the local trend than it is to hunt out a plan that is suitable for individual needs. In offering this barn plan it is with the idea that there are many locations where the style of building and the shape of the roof will match the house and other surroundings better than any other plan.

A roof like this is not economical to build if the owner is influenced especially by dollars and cents, but there is a style some way so it is not obtrusive. But there is something wrong with a man who will build a gothic house and a barn with a flat roof on the same lot. His ideas have been dwarfed in some direction. His property shows it because it does not balance up right.

A lot with its buildings must be one homogenous whole or it shows at once that it has not been arranged rightly. A

FRONT ELEVATION SIDE ELEVATION

DESIGN FOR A SMALL BARN WITH FOUR STALLS

about it that shows up well for the amount of money it costs. There is a great deal in appearance. When we have things right we have something to appreciate for a long time to come. If the house has a steep roof we cannot tolerate a barn with a main roof that is, say one-third pitch and a lean-to that is even less.

If the mischief has been done conditions may be somewhat improved by moving the barn back well out of the way and having it covered with vines or screened in village stable may be made an ornament to the property or a damage to the owner and an eyesore in the neighborhood. Neighbors often say unkind things about the owner of the barn on the next lot. Not always on account of the looks of the thing; they may be aggravated by the perfume or the noise of the chickens when they want to sleep in the morning.

A good many folks don't like neighbors and it is generally for some such reason, but neighbors are necessary and the neighbors sometimes build barns and they don't always keep them nicely. It requires a level-headed man to lay out a lot to the best advantage and put up buildings in such a way that no one can find fault with them.

There is something about the arrangement of this barn inside that will appeal to every orderly person. The stalls are right for convenience both in handling the horses and for cleaning the stable. The carriage room is quite large and convenient with two store rooms, one for general garden tools with a place for small work bench on one side, a necessity in almost any village lot where a man is kept to do the chores. The other storeroom is intended for harness. There is also a case which comes in very handy to keep the smaller things and those that are valuable. The glass doors slide past each other and may be easily locked shut. It is a good plan to have some little cupboard like this that may be locked when occasion requires it. In almost every stable medicines are kept and they should be out of the way of children. It is a splendid precaution to keep medicine bottles locked up. A great many accidents have come just from carelessness in this respect.

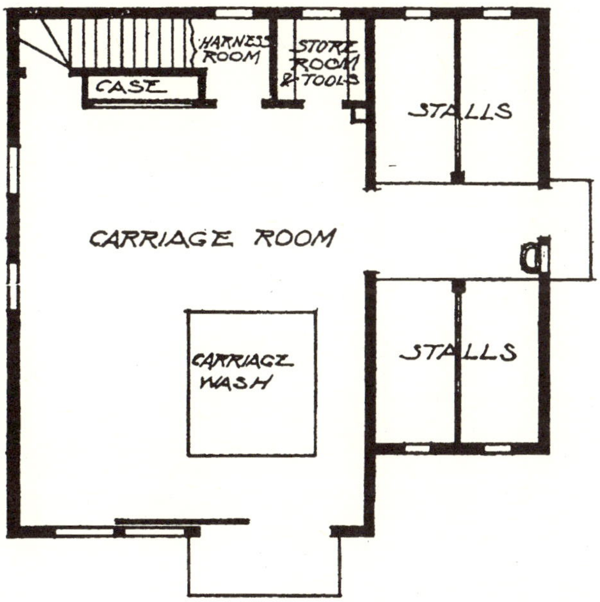

FIRST FLOOR PLAN

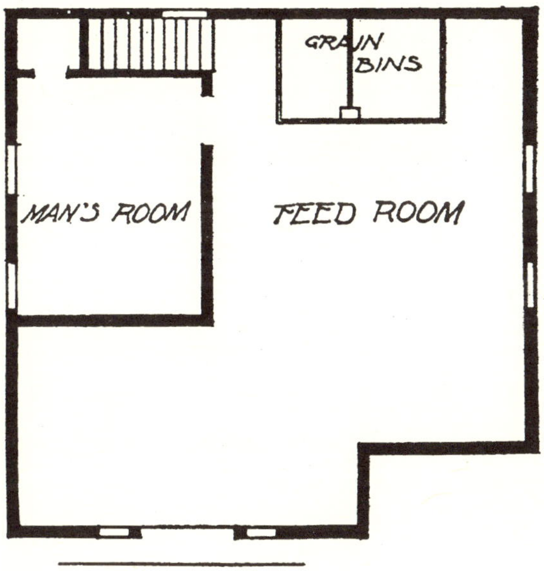

SECOND FLOOR

Every village stable that is large enough should have a room for the man; it may not be necessary at all times, but the time will probably come when this room will be found very useful. In this case it is built in one of the large gables where the roof is steep enough to lath and plaster right on the rafters. It is a case of building a roof and a side at the same time and it makes a saving in expense in one way or the other. You either don't pay for the roof or you don't pay for the side of the room.

CEMENT ROUGH CAST BARN— A182

Cost of Blue Prints, $15.00

A carriage house and stable plastered on the outside with cement mortar with a rough cast finish is shown in plan (No. A182). There are locations where a basement for laundry purposes under the house is not desirable. This plan for a carriage house with a laundry attachment was designed especially to meet such cases. In

New Orleans, La., such carriage houses are quite common. There is a great deal of made ground and the sewers are not deep enough to permit much underground building, so that basement laundries are not common. To meet just such conditions stables with laundry rooms just seem to fill the bill, especially when they are well designed and built to suit individual needs.

This building is substantial in appearance and the manner of construction is very satisfactory for a warm climate. The outside cement work when properly put on with metal lath is very durable. It looks well and is not expensive.

The usual conveniences found in small barns are provided in this building, but it is a little more elaborate than ordinary. The box stalls are especially large and

FRONT ELEVATION

REAR ELEVATION

roomy, there is a larger feed room than is customary and the harness room is a little larger than we usually find in a small

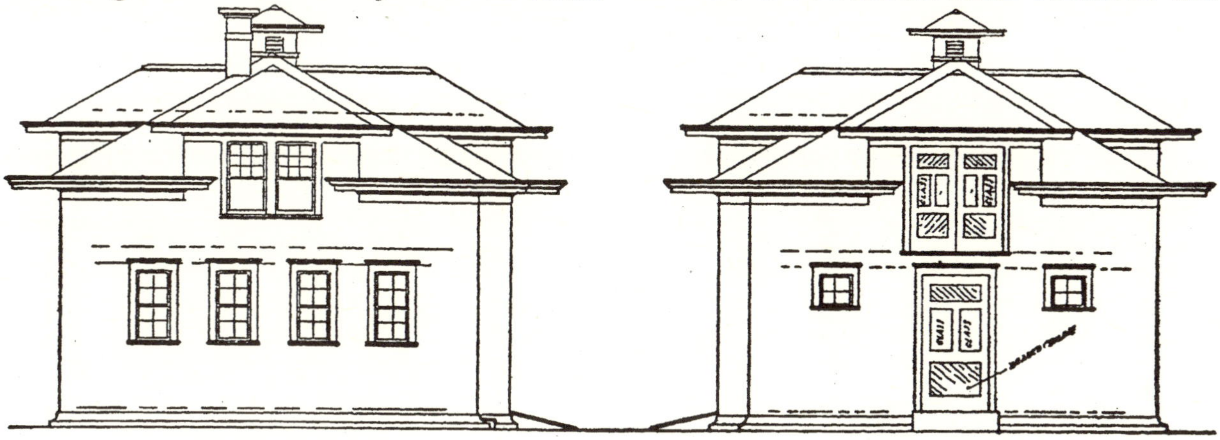

LEFT SIDE ELEVATION   RIGHT SIDE ELEVATION

or medium sized stable. But the especial features about the building are the rooms

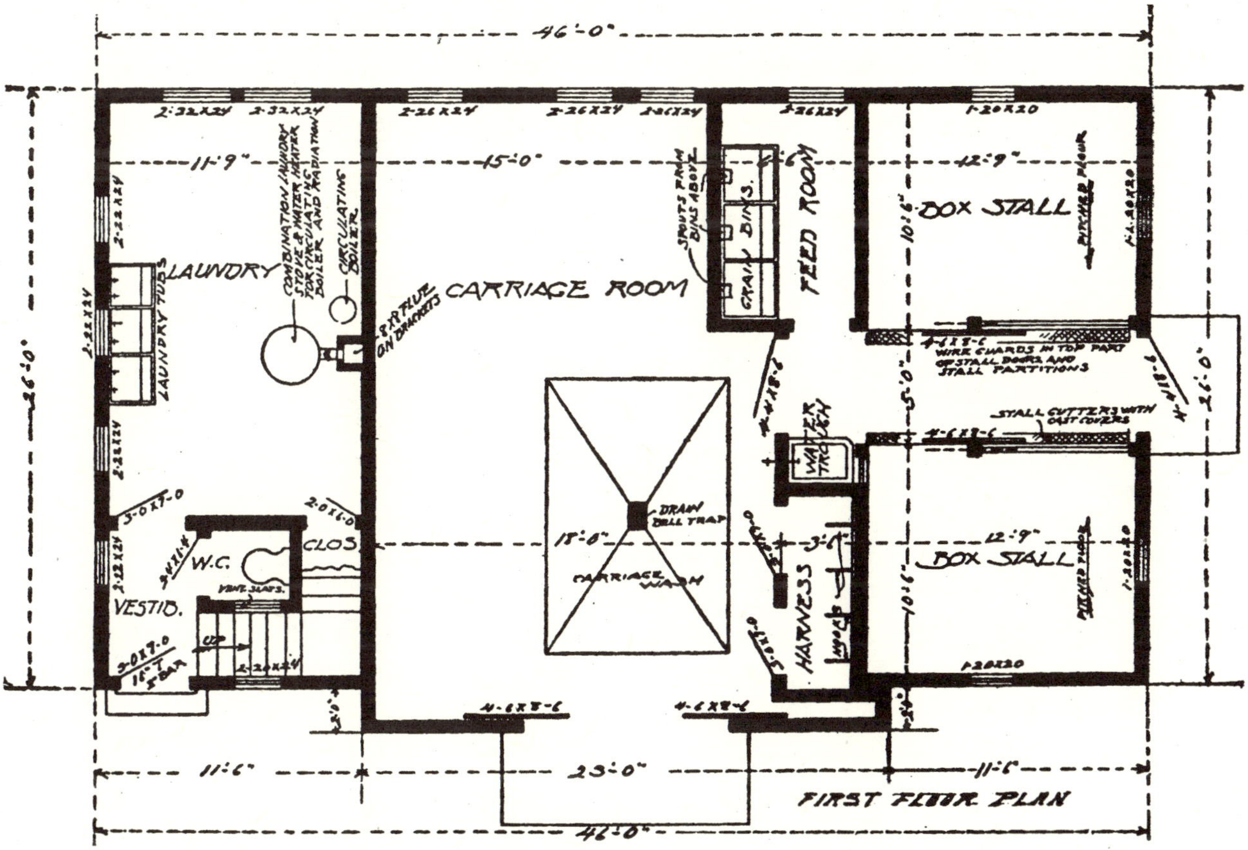

for servants with an entrance separate from the carriage house, and the laundry with its hot water heating apparatus, which not only furnishes hot water for washing and for stable use, but to warm the stables and the servants' rooms in winter. This laundry room is also large enough to hold the clothes lines in stormy weather, and there are plenty of windows for light.

Laundry work is a problem in the south

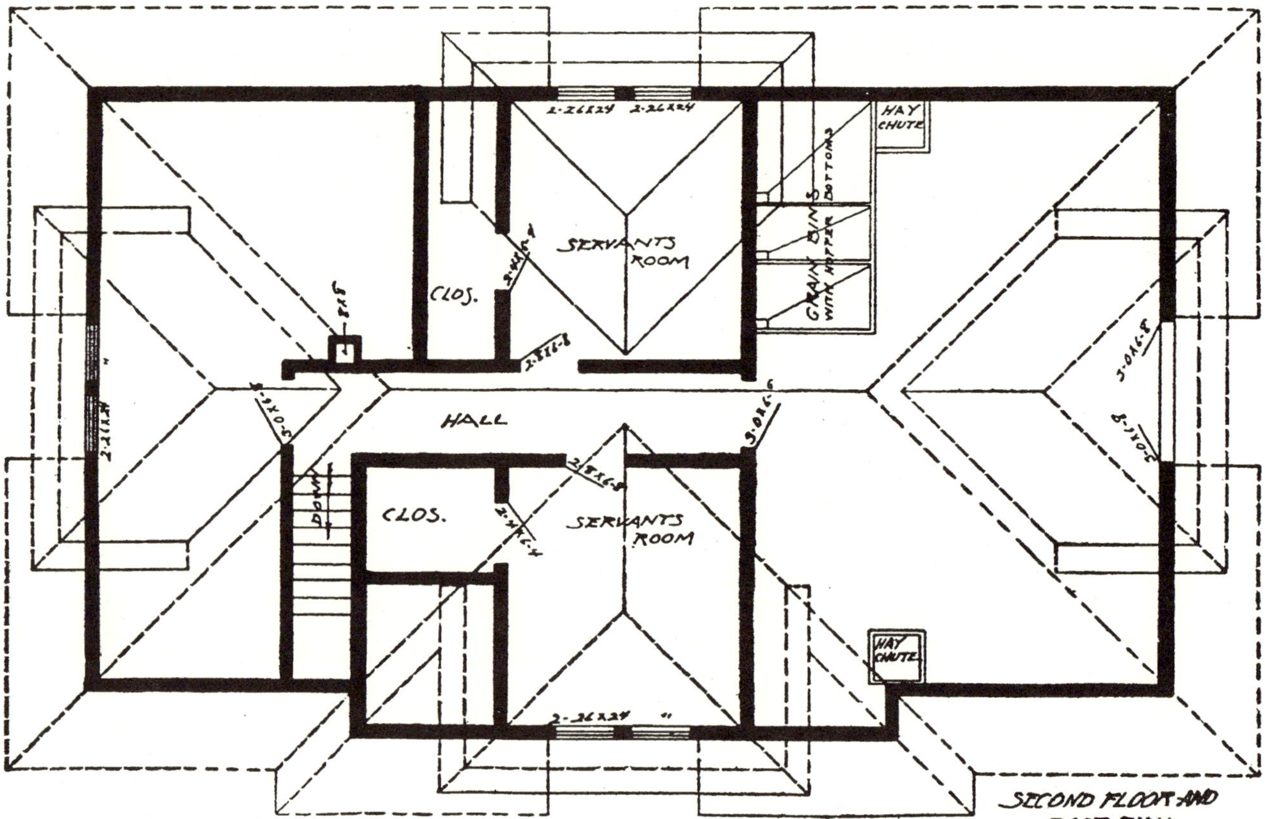

SECOND FLOOR AND ROOF PLAN

as well as in the north. Those who get along with the least friction usually have the best possible conveniences for doing the work. Large light laundry rooms supplied with plenty of hot water and furnished with good machinery and tubs that are rightly placed and fitted with the necessary faucets, waste pipes, etc., offer more inducements to do good work and less occasion for complaints than ordinary.

There are many advantages in having the laundry room away from the house. It avoids confusion in the house on wash days and the odors of dirty steam and soapy water are done away with.

For a pretentious property a stable building of this design and size looks well. The building is large enough to match up well with a good big residence and the design and style of the roof shows character enough for a house, in fact many costly houses are built with roofs that are less attractive than this one. A carriage house like this is not complete without a good wide drive leading to it. This design requires a smooth pavement in front of the building one-third wider than the building itself. It should have a pretentious approach to give it proper setting. Sometimes an inferior building can be given a royal appearance by an elaborate entrance. A driveway to the stable is part of the entrance. In this plan the inside is right, the outside looks well.

HORSE AND COW HOUSE—A131

Cost of Blue Prints, $5.00

A small carriage house with stable room for two horses or a horse and a cow is a very convenient thing when a person has a good sized lot in the city or village. A horse stall makes a splendid stall for a cow, better than what is ordinarily designed for a cow stall because there is more room and it gives more comfort. A cow appreciates comfort and will give

FRONT ELEVATION

enough more milk to pay for it. Of course a cow in a horse stall needs plenty of bedding, but where only one cow is kept it is easy enough to furnish all the litter necessary.

There are a good many designs for small

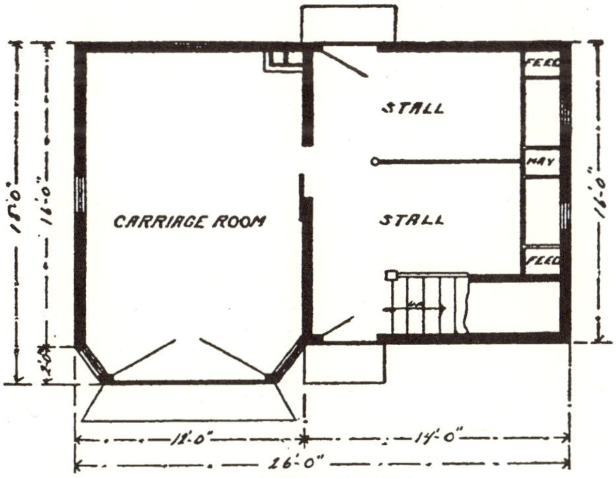

FLOOR PLAN

carriage houses, some of which are decidedly homely. A good many of the fancy buildings are too expensive. Here is a comparatively cheap structure, but it is all right for looks and it is a convenient stable to do work in. There is a hay chute which reaches from the loft to the manger below with openings for both stalls, which is a very convenient arrangement and is

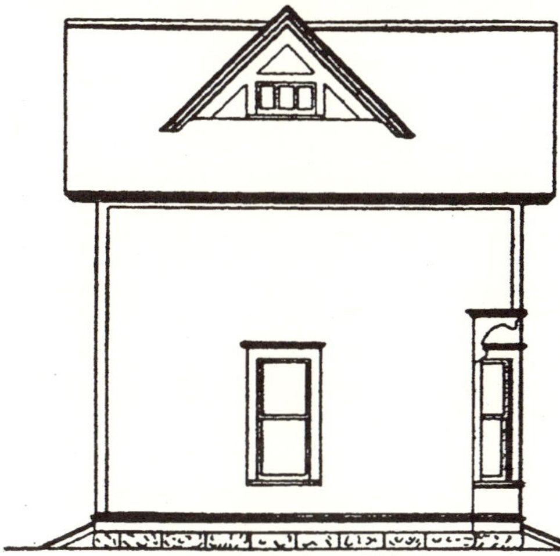

SIDE ELEVATION

worth a good deal just to keep the hay dust and chaff out of the horse's mane and fore tcp. It also leaves the feed boxes in the corner of the mangers for grain and other feeds.

A carriage house like this may have a

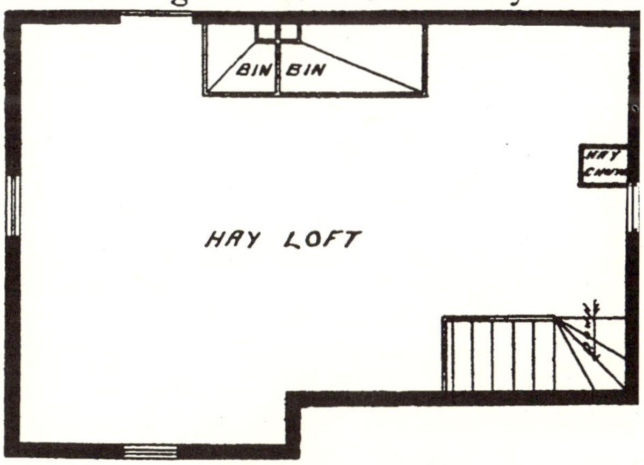

SECOND FLOOR

plank floor or the floor may be left out entierly and the ground leveled up with cinders except the stalls and the very best stall floor is made of stiff clay pounded in wet. Some of the most successful horsemen prefer a clay bottomed stall.

CARRIAGE HOUSE AND STABLE—A127

Cost of Blue Prints, $5.00

The illustration on next page shows a carriage house and stable twenty by thirty feet on the ground and fourteen feet high to the plates. The ceiling is eight feet six inches which is about as low as you can have a ceiling in a carriage house because really requires about three times as much room to stable horses this way.

FRONT ELEVATION.

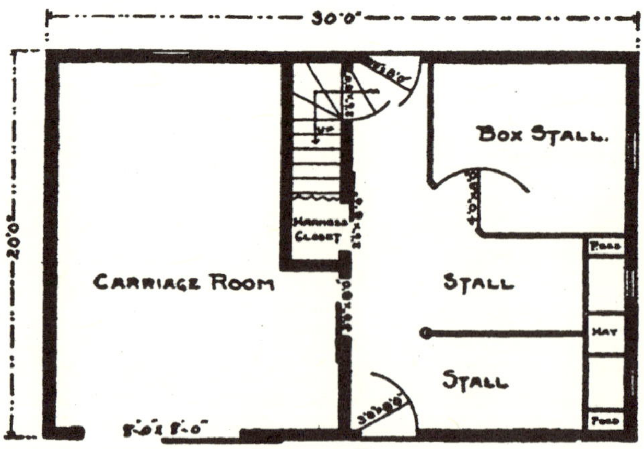

FLOOR PLAN.

you must have room enough for a top buggy. For this reason the doorway must be about the same height.

The internal arrangement of this stable is different from most small carriage houses. There is a box stall about nine feet square. It is difficult to plan a decent sized box stall in a small stable. They run into room too fast. Nothing looks so comfortable for a good horse as a roomy box stall. If the horses had their say about it there would be more box stalls, but it

HORSE SHED—A121

On farms where a number of brood mares are kept and colts of all ages coming along, it is much better to have a separate shed for winter feeding for the colts than let them run at large among the cattle. One colt might not do much damage in the general barnyard, but colts are mischievous and one teaches another.

A light shed may be built on this plan, which is fifteen by thirty-four feet, at very little expense. It should front on the stack

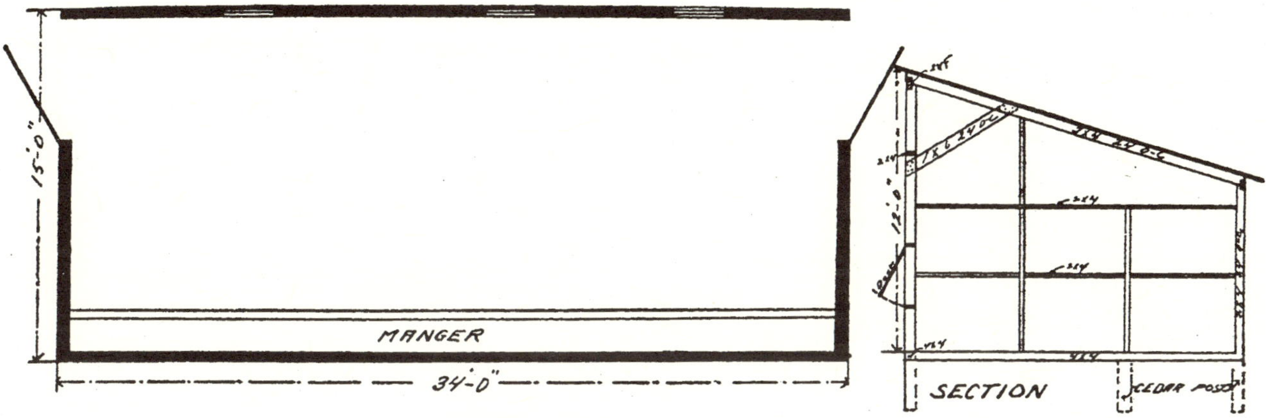

HORSE SHED

yard and face the south if possible. For economy it is placed on cedar posts let in the ground below frost, but it should be thoroughly banked up in the fall to keep out the cold winds. In banking up a shed like this set a board all around the outside to keep the earth away from the building proper. Fit the board nicely so there are no chinks to let in the cold draft.

WELL PLANNED HORSE BARN—A171

Cost of Blue Prints, $10.00

To make this article more comprehensive to those interested in barn construction we show an exact reproduction of

NORTH ELEVATION OF HORSE BARN

the architectural plans after which the building was erected.

This building is designed to accommodate fourteen horses, having ten single stalls and four box stalls, and all the necessary feed bins, harness room, wash room, grain room, carriage room, storage rooms,etc.

The carriage room, which is 30 by 36 feet clear span without posts, is on the east end and has an entrance of large double sliding doors, and also a large sliding door to the horse stable. The carriage room floor contains a carriage wash near its center and overhead is a large trap door, so any vehicles which are out of use

can be hoisted up to the floor above for storage. The carriage room also has direct doors to the harness washing room.

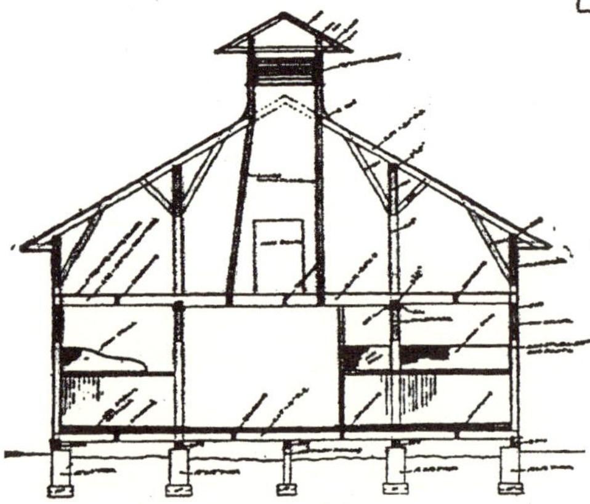

CROSS-SECTION

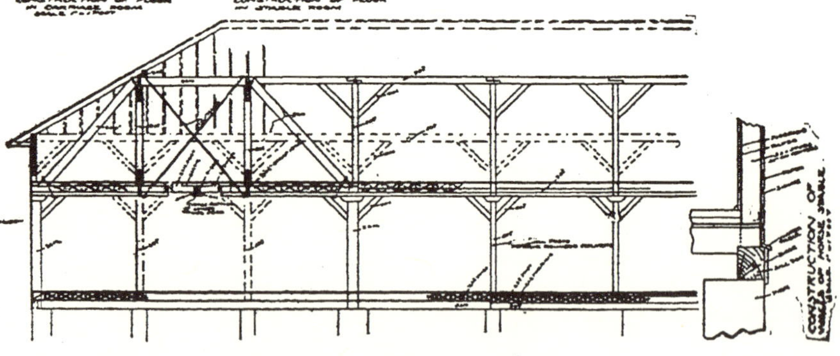

LONGITUDINAL SECTION OF FRAMING

The harness room is equipped with dust proof cases for the harness, blankets, etc., and the washing room contains a sink with soft water supply and all the necessary fixtures required for the washing and repairing of the harness.

The stable room contains a watering trough, a store room for tools, shovels, etc., and a grain room for the mixing of feed, and which has small grain bins which are connected with spouts from the larger bins on the upper floor.

The box stalls have sliding doors with a wire grill in the top half, and the partitions between all stalls have wire grills running up to a height of about 7 feet above the floor, thus obtining a free circulation of light and air. Each stall is equipped with a window that is hinged on top and swinging out. This provides each animal with fresh air and a direct draft upon the animal is avoided by these windows being placed up near the ceiling, also being covered with a wire screen for protection. All stalls have cast iron feed boxes, salt boxes and wrought iron hay racks connected directly with hay chutes from the hay room above. All stall floors are slightly sloped to the back and there connected with a cast iron drain trough running the full length of and on each side of the driveway.

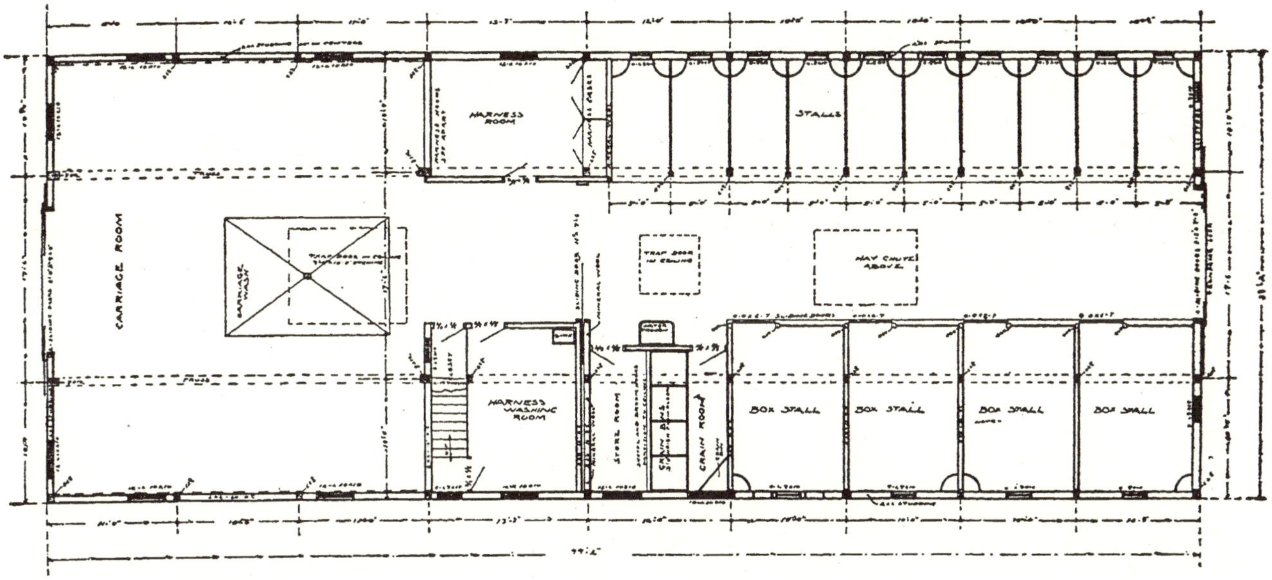

In the ceiling of this driveway is a large trap door for throwing down hay and bedding, and also for the hoisting of hay from the hay wagon in stormy weather.

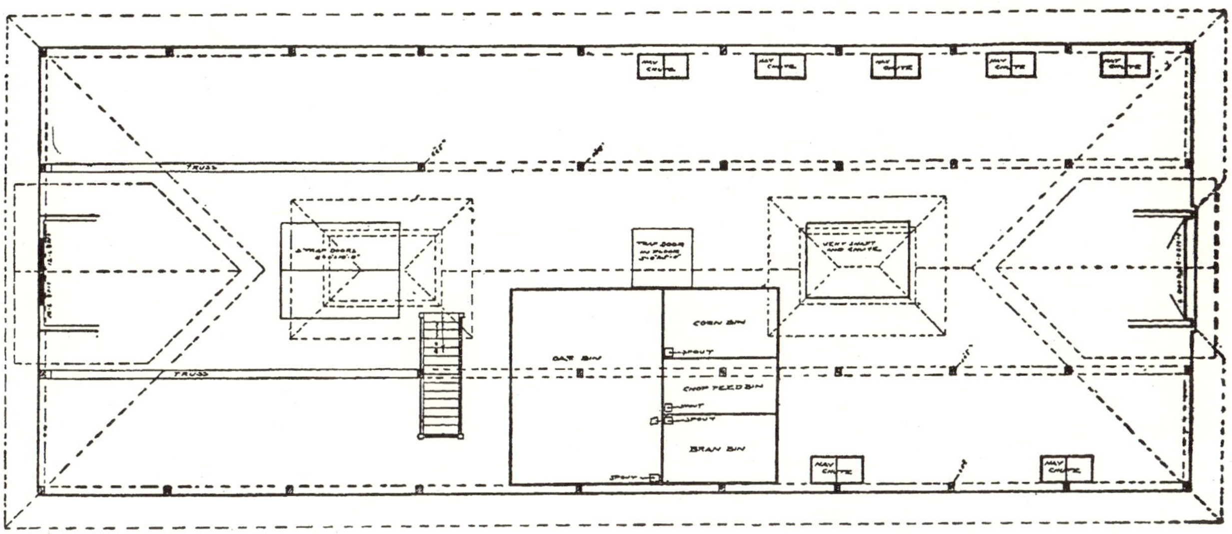

One of the roof ventilators has a shaft running down to the ceiling of the horse stable for ventilation, and is at this ceiling provided with trap doors by which the flow of air can be regulated as desired, and this shaft at the same time serving for a hay and bedding chute.

The second story is used for the storage of hay, bedding, grain and feed, and the room above the carriage room is partitioned off into a dust-proof room for the storage of vehicles, etc.

This building is built on a foundation of stone piers, so as to admit a free circulation of air under the floor and to prevent the floor from becoming cold in the horse stable it is built, as will be seen in the detail above the longitudinal section, by first resting the joists upon the sills, then floored with a matched floor 1 inch thick, which is covered with a heavy building paper, then by 2 by 2 inch strips nailed one over each joist. The space between these strips is filled with mineral wool, then this entire surface is floored with a strong floor 1¾ inches thick, and on this are laid strips of various thickness to receive and form a pitch to the stall floors. On the sills over each stone pier is set a 6 by 8 inch post for the support of the second story floor and roof. These posts run up to the plate, which is a 6 by 8 inch timber, and at the second story joist level there is a 6 by 8 inch timber notched in between these posts for the bearing plate of the second floor joist. All these timbers are braced at all intersections with 4 by 6 inch braces. The outside walls are formed by filling in between these bents with 2 by 6 inch studding spaced 2 feet on centers and well spiked to the floor joist, sills and plates. The inside surface of these studding are covered with heavy building paper, then ceiled with matched flooring, and the outside surface of studding is also covered with paper and then sided with drop siding. The roof is of cedar shingles dipped in moss-green creosote stain, which in contrast with the white painted walls, makes a very artistic effect. The interior of the carriage room is finished in yellow pine beaded ceiling.

RESIDENCE BARN—A216

Cost of Blue Prints, $15.00

A residence barn to accommodate three vehicles and three horses. The carriage room is of good proportions and has a wide door at front and rear. The harness room and man's room are of good size, and conveniently located.

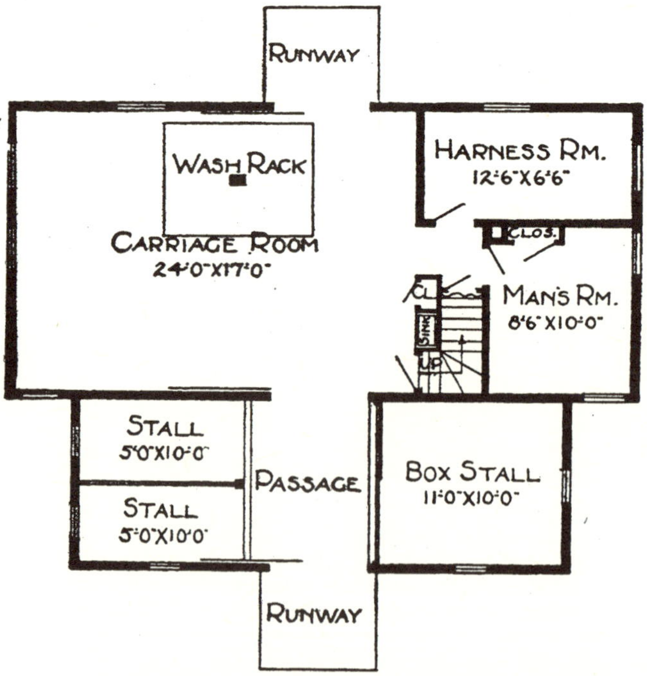

The construction of this barn is of the balloon frame; there is a concrete foundation. The ground floor is of cement and all rooms are cement wainscoted up to the window sills, making the walls waterproof.

The exterior design is of a modern style with a Japanese style of roof which gives the building a very odd but charming artistic appearance.

There are so many windows in this stable that it is very light inside and they give it an expensive look outside. It is not generally recognized that windows are about as cheap nowadays as any other part of the building. As soon as people generally grasp this idea all farm buildings will be made lighter, more sanitary and more cheerful.

The side walls are rough cast cement up to the windows and the balance of exterior vertical walls sided. The roof is of moss green stained shingle, which in connection with the white siding, grey cement and brown stained trimmings, makes a very striking exterior that would do credit to any neighborhood where the commonplace board-and-batten barn would be objectionable. This barn, though somewhat artistic in its outline, can be built at a reasonable price, and contains no work that cannot be executed with materials that can be bought from the stock of the lumberyard.

The interior makes very good provision for two or for three horses, there being two single stalls and a large box stall. The man's room is well finished and is very pleasant. The harness room is large and nicely lighted. There is a large convenient loft for hay and grain storage. The entire barn is exceptionally well lighted and ventilated.

SMALL LIVERY BARN—A138

For a village or a small city this plan offers a comparatively cheap building that may be used to advantage by a man who keeps four or five horses for hire. Usually in such cases it is not necessary to have a great deal of feed storage room because the hay is baled and sometimes the straw comes in bales. A good harness room is necessary and it often happens that the hostler wants to sleep in the stable and this room, ten by fifteen feet, is sufficient for such purposes.

FRONT ELEVATION

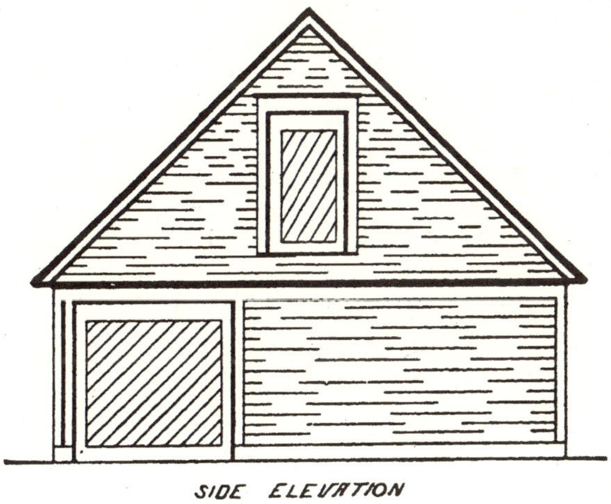

SIDE ELEVATION

The problem in all livery stables is how to take care of the different rigs. There are cutters and sleighs to be taken care of nine or ten months in the year, when they are not in use, and there are wagons in the way almost all the time. Storage room is expensive and sometimes ground room is an object.

Too often public stables are littered around outside of the building with old trash that should be sold for junk or burned up. Such conditions are more noticeable in the smaller places. But pride in keeping up one's property is just as valuable and just as necessary in a village as in the city. Perhaps liverymen and blacksmiths are a little more careless in this respect than any other class of citizens. Why this should be so is a mystery. It costs nothing to be neat and neatness attracts trade in these lines as well as others. From general observation it would seem that a place for everything and everything in its place is a suggestion which applies to liverymen and blacksmiths all over the country.

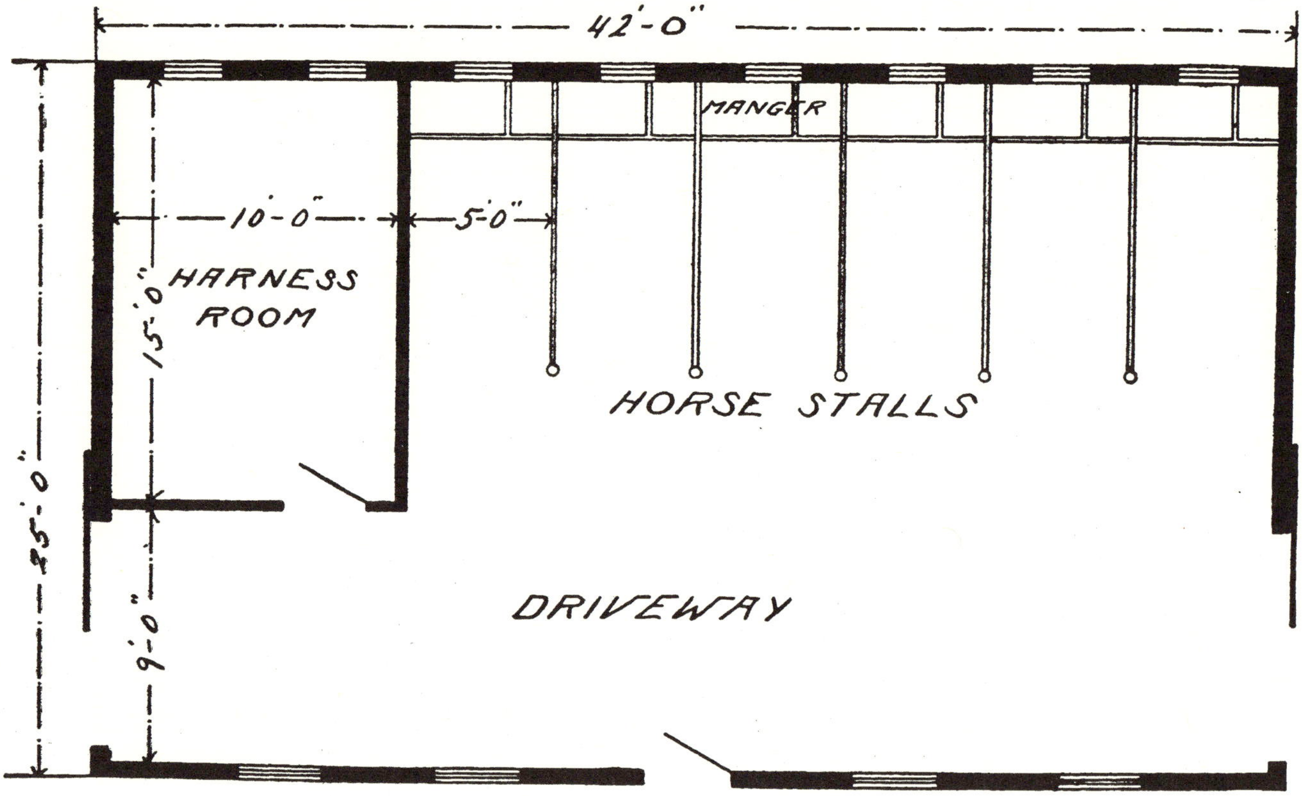

FLOOR PLAN OF SMALL LIVERY

LIVERY STABLE—A218

This stable is of frame construction, built on concrete foundation and has concrete floor throughout the entire ground floor area. The front building, containing the office, carriage room, wash room, etc., is separated from the building containing the horses by a cement fire-proof wall and fire-proof door. The entire exterior wall surface of building is covered with galvanized iron siding and a corrugated iron roof, making the exterior practically proof against fire. The boiler room has brick walls, fire-proof doors and cement floor and ceiling. With these precautions against fire, electric light being used for illumination, the building is reasonably safe, although built with wood walls.

The stable contains thirty stalls, one of which is a box stall for sick horses, with double doors from the yard and single door from driveway. All stalls have removable plank floors laid on the cement floors, with slight pitch toward the rear of stalls. The stall partitions are of matched plank, four feet high and have an iron guard on top, making top of guard seven feet above floor. Each stall is provided with a hay manger and feed box. There is a hay chute of galvanized iron between each two stalls, this hay chute running from second story floor to top of hay manger; is built larger at the bottom than at the top to prevent hay from clogging. The bottom of mangers and feed boxes are filled with at least four inches of cement to prevent the horses from biting into the planks, and the front edge of the manger is covered with strap iron fastened with counter-sunk screws.

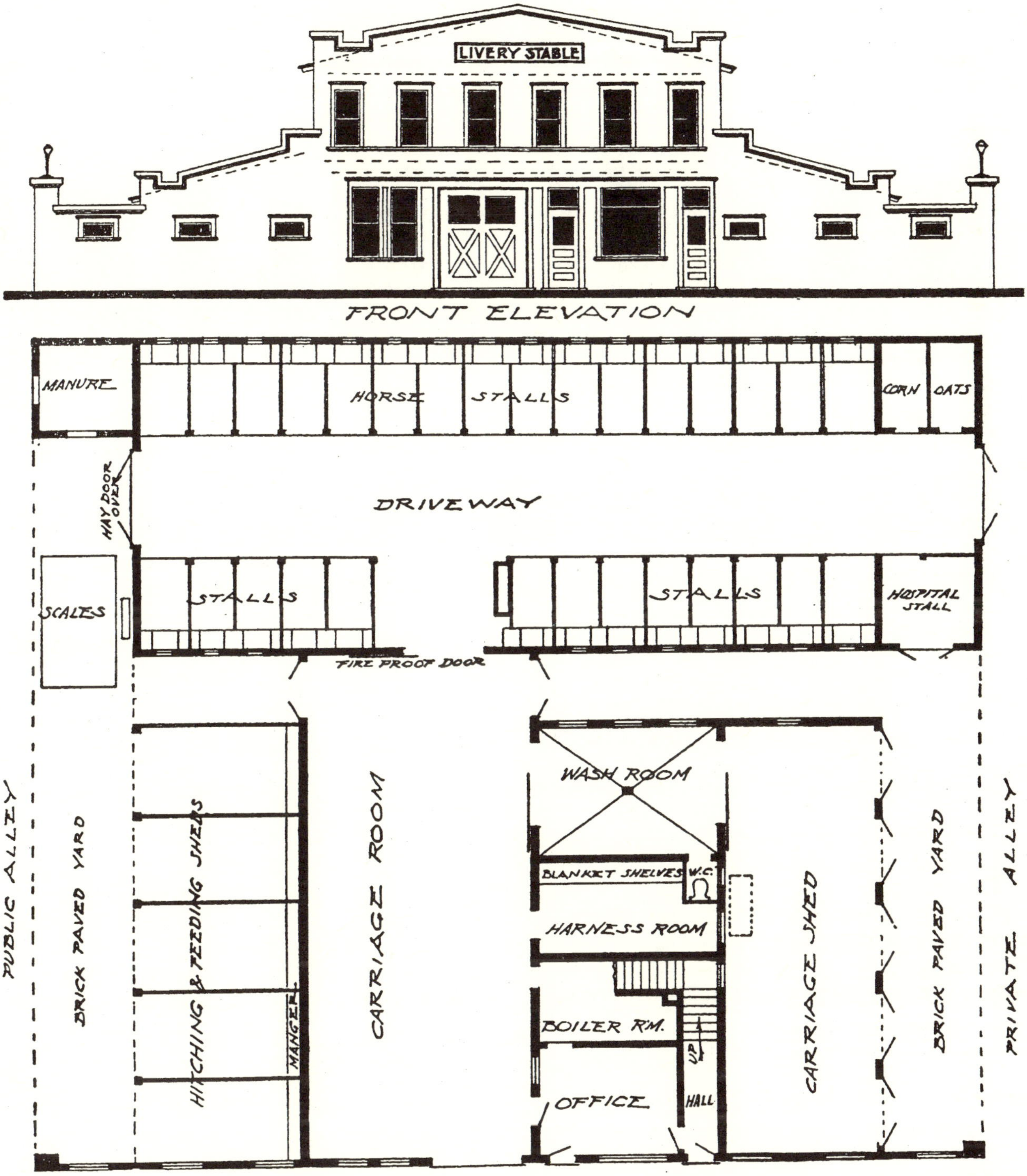

The carriage room has a clear 22 foot span and is 50 feet long. At the rear it contains a vehicle hoist or elevator to additional storage room and paint shop, located at the rear end of second floor over carriage room. The front end of second floor facing the street, contains a comfortable flat, consisting of a parlor, living room, dining room, pantry, kitchen, bathroom and three bed rooms of good size, all rooms having outside windows and good ventilation. The second story over the horse stable contains a large hay room, bedding room, storage room, and grain bins.

PITCH OF BARN ROOFS—A228

One-third pitch means that the peak of a roof is about one-third of the width of the building higher than the plates, that is if the building is thirty feet wide the peak has an elevation of 9 feet, scale measure above the plates. Half-pitch would be half the diameter higher, or fifteen feet above the plates, while full pitch would be 26 feet scale measure.

What interests a farmer most in the pitch of a barn roof is the storage capacity as compared with the expense. Any pitch from one-third up is a good one so far as service and lasting qualities go. A half-pitch might last a little longer than a third-pitch, but there is not enough difference to pay for the extra cost. Full pitch is used only for architectural effect except in cases of gambrel roofs when half pitch or even steeper is often used for the lower portion between the gambrel and the eaves.

In gambrel roofs the upper section often is as flat as one-fourth pitch while the distance from the curb to the eaves is sometimes very steep and both sections are short.

Some farmers claim that it is just about as cheap to carry the sides of a barn a few feet higher and use a plain one-third pitch roof while others feel that it is a great deal better to set the plates at the usual height and carry the roof up in gambrel roof form.

It depends somewhat on the size of the building and proportions, and the length of material that can be secured. Twelve and sixteen foot boarding are the commonest lengths. The higher you go above the plates the more expensive is the building in proportion, because scaffolding must be higher and all work above a certain height costs more. The work on the roof is a little different because the roof is

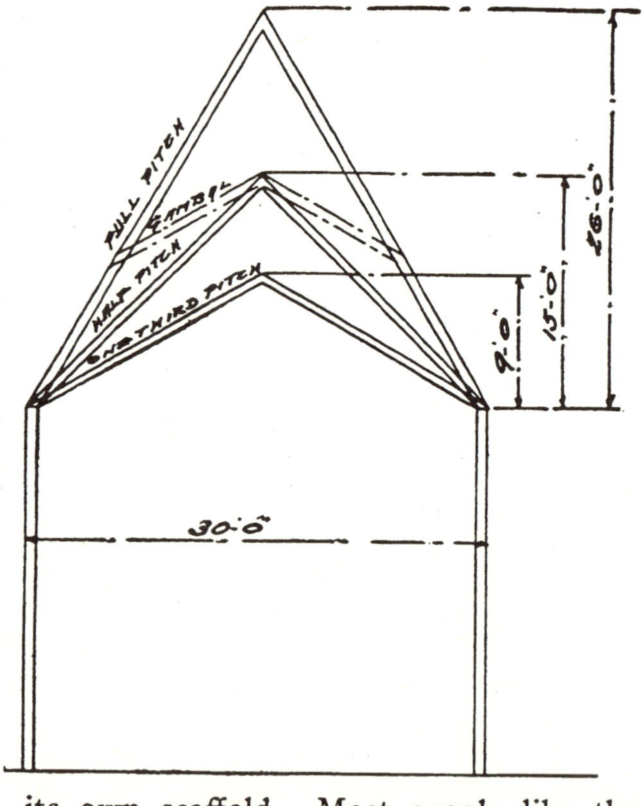

its own scaffold. Most people like the looks of a gambrel roof on a barn, and looks count for a good deal after the building is finished.

Feed Lots and Cattle Sheds

FEED LOTS FOR BEEF CATTLE—A184

WHERE cattle are fed in large numbers it pays and pays well to fit up properly for the business. In the corn belt, buying thrifty young cattle and finishing them for the market, is a splendid business in the hands of men who understand how to buy, how to feed and how to sell. The old fashioned way of putting a fence around a mud-hole and confining a bunch of cattle in the mire for weeks or months at a time ceased to be profitable long ago, but unfortunately some men haven't found it out. Considerable engineering ability is required to plan and construct feed lots for the accommodation of large numbers of cattle in such a way as to make the animals comfortable and to economize labor.

Plan (A184) has received very careful attention in this respect. The storage barn and silos are set on a ridge of ground sloping preferably to the southwest. The feed lots, thirty-two by seventy-two feet in size, including the shed, are fenced off one after another as many as needed. Two yards only are shown in the drawings because no matter how many you have each pair of two would be a repetition of this pair. The lots might be extended a quarter of mile holding the same order.

It works better if the ground is about eight feet lower for the feed lots than it is for the storage barn and silos as this gives a chance to run the track from the floor of the storage barn over the heads of the cattle high enough to leave a passageway under for a pair of horses and a manure spreader. Eight feet in the clear is little enough and it is high enough because straw as well as feed will be brought to each lot by a car on the overhead track. The car is made large for this purpose, being four feet wide at the bottom, six feet wide at the top, four feet high and eight feet long. When filled with silage it will make quite a load, but one man can move it if the wheels are large and kept well oiled and if the truck is level and true. Some feeding yards have an incline track, but this is not necessary, in fact it is objectionable because the car will never stay where you want it and it is uphill work getting it back to be refilled. Make the track absolutely dead level and perfectly straight. Two by fours plated on top with two inch band iron that has been hammered straight and true will answer very well but the two by fours must be well supported and thoroughly spiked in place. In building the track remember that you are trying to save time and labor at every feeding period for a number of years to come. You want the track so true and the car wheels to fit so perfectly that the car will run along without much friction after getting started.

One man with a rig like this that works right should feed a large bunch of cattle because he can take advantage of his work.

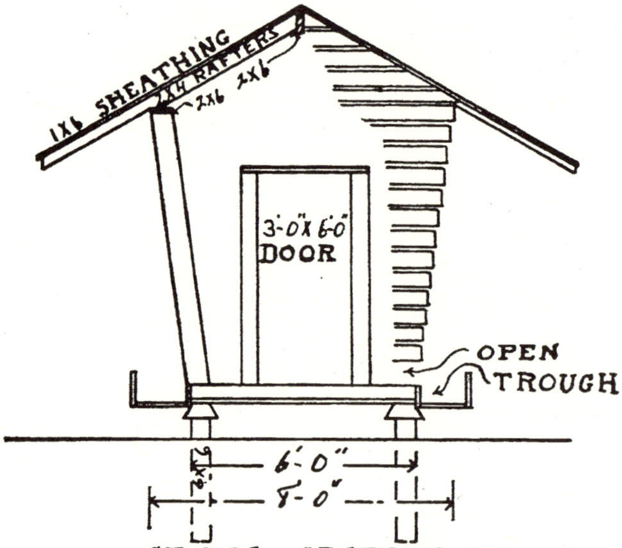

CROSS SECTION OF
CORN CRIB FEEDER

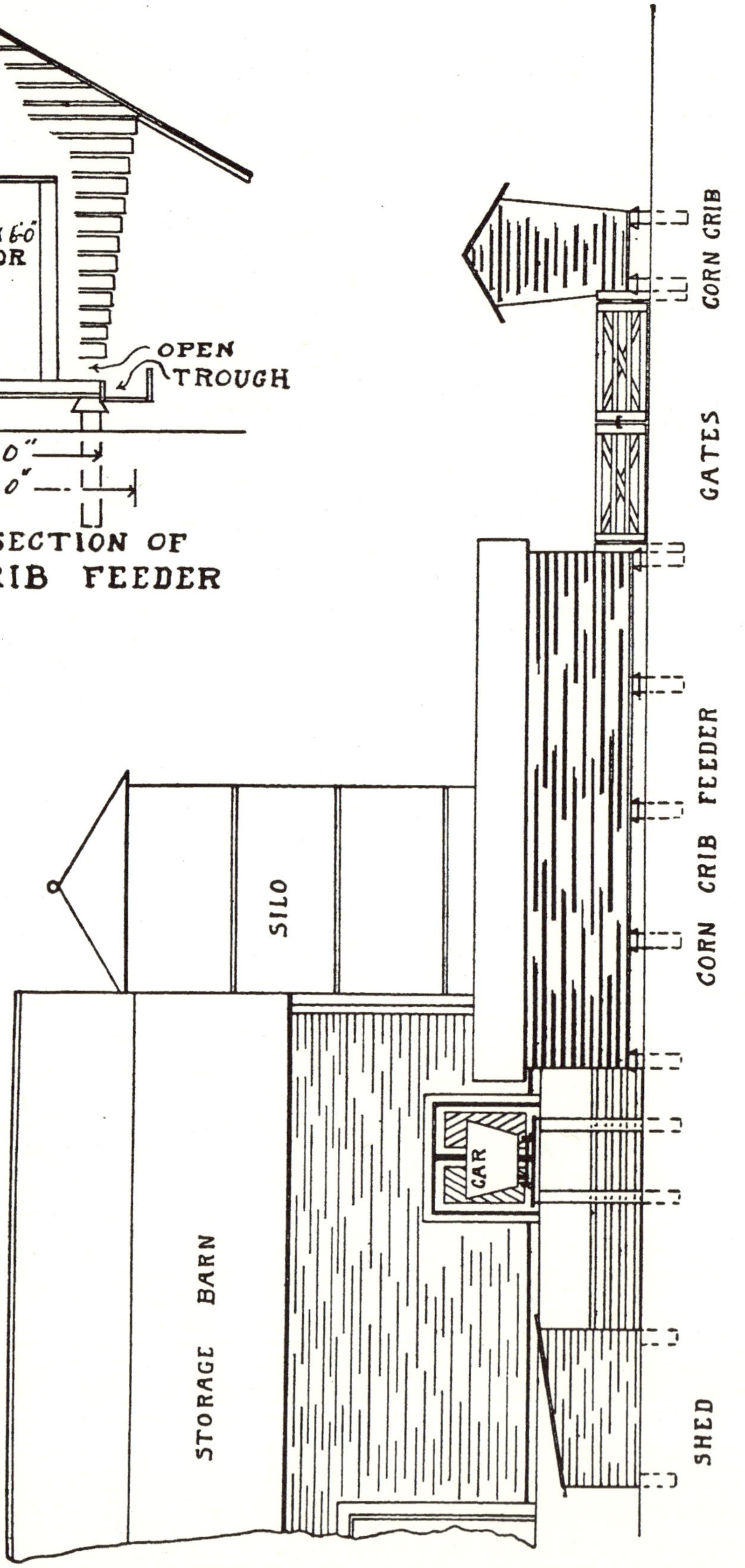

WEST SIDE ELEVATION

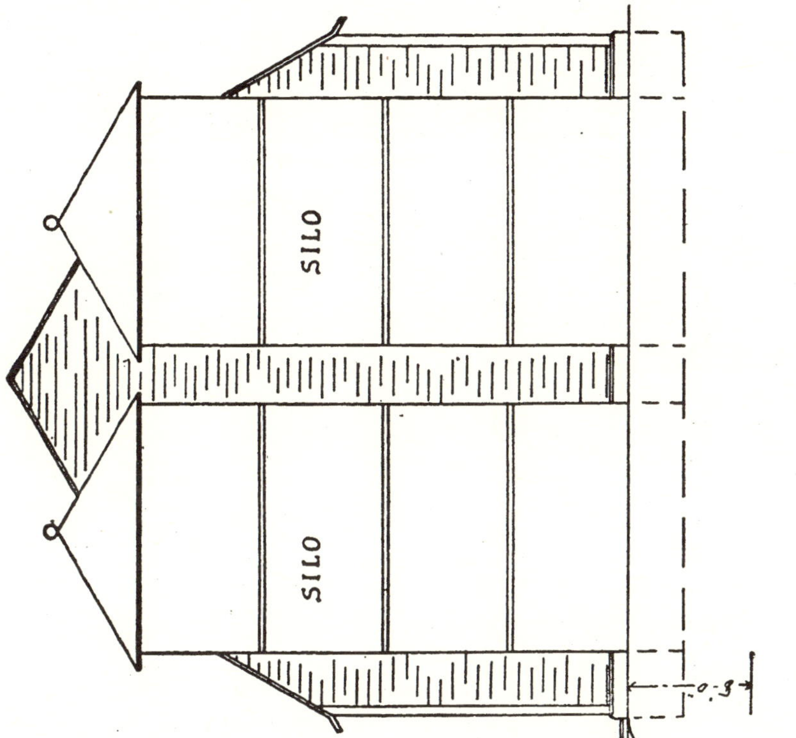

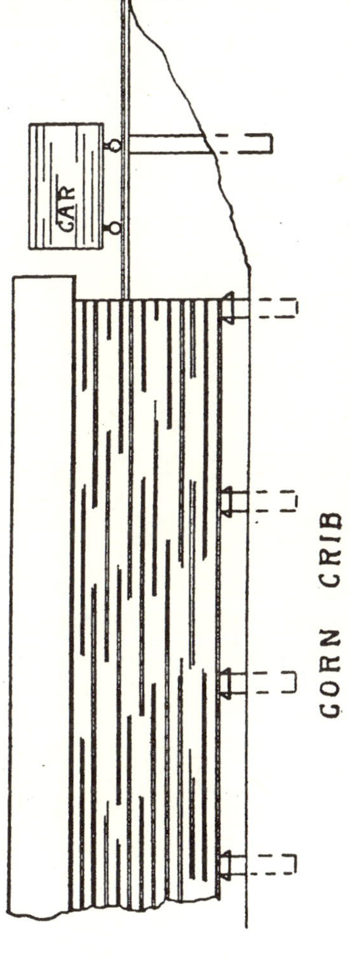

SOUTH SIDE ELEVATION

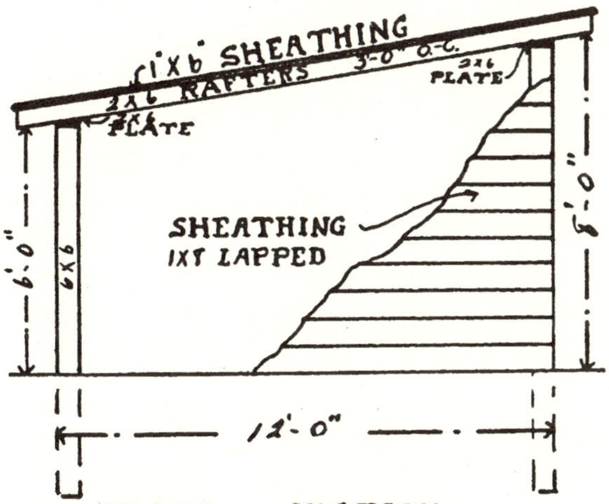

CROSS SECTION OF SHED

In the first place he has got a car big enough to hold something. He runs a chute from the silo to the car which saves forking the silage up from the floor until the silo is nearly empty. The sides of the car are hinged so they drop down over the feeding racks in the yards. He loads the car quickly and easily and a good deal of the stuff unloads itself. The track is made in sixteen foot sections, as the yards are thirty-two feet wide the tracks have one support in the middle of the yard. The other supports form part of the fences between the yards.

In laying out the yards the problem of drainage must be worked out first. It is impossible to have the yards dry unless ample provision is made for taking care of the rainfall. A drain tile is marked on the plan leading from the corner of the storage barn and running across the ends of the feeding pens down the whole length of the alley to an outlet in the field beyond. The brick pavement in each feed lot slopes to the center to lead the water to the tile drain underneath which connects with the trunk line of tile near the fence in the alley. This main drain increases in size to accommodate the extra drainage as it proceeds past the different pens.

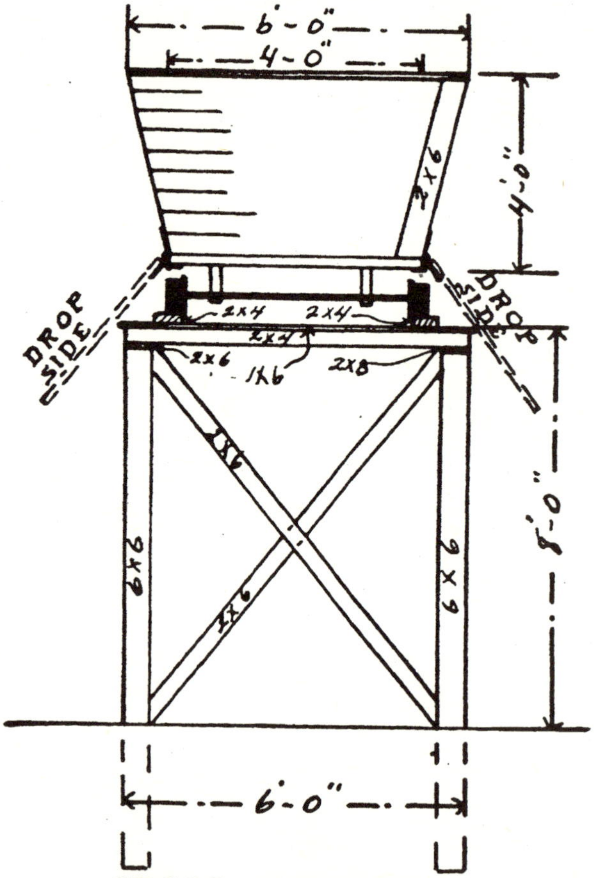

CROSS SECTION OF CAR

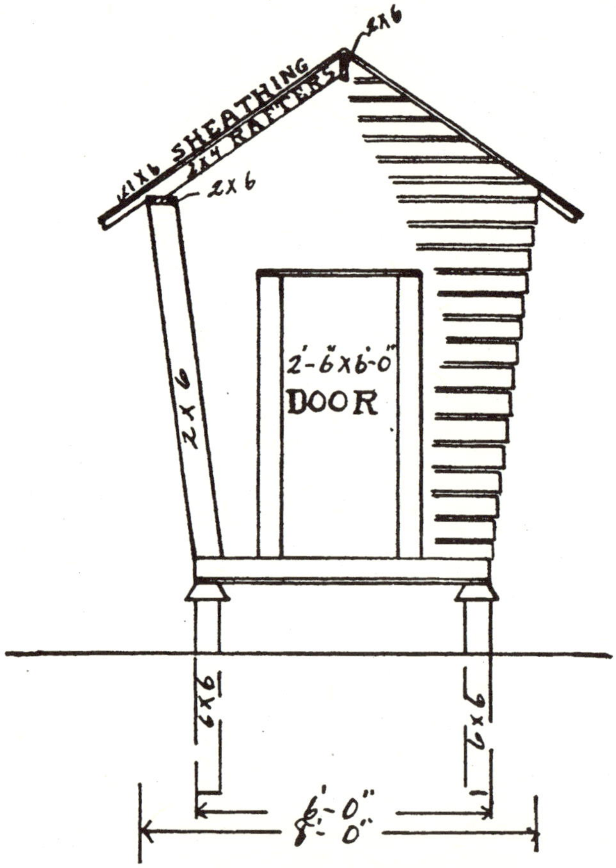

CROSS SECTION OF CORN CRIB

An open shed twelve by thirty-two feet occupies one end of each yard. This shed is not paved but is kept well bedded. All the rest of the yard is paved with brick laid flat on a cinder bed.

An additional drain tile runs from each water tank to the trunk tile line to take care of any overflow from the tank. In some locations another tile drain will be necessary at the back of the shed because the ground must be kept dry.

Water Supply.

Good fresh water in sufficient quantity to supply the needs of the cattle in these feed lots is quite a problem in itself. The water must be good and there must be plenty of it. It must also be supplied under pressure and carried to each water tank in pipes placed under ground below frost. There must be a valve placed in each pipe running to each water tank so constructed that it won't freeze. The stems from these valves should be extended up to the overhead track so a man can walk from one end of the feeding yards to the other and regulate the water easily and quickly.

Generally the water must be supplied by a windmill and a reservoir of some kind. A cement basin in a nearby hillside is perhaps the most satisfactory because when once made it is permanent. The source must be sufficient to supply it and the windmill or other power which does the pumping must be powerful enough to do the work at all times. You cannot afford to take chances on a water famine with several hundreds of feeding cattle on your hands.

Storage Barn.

In the plan not much attention is paid to the storage barn except that it shows the most convenient location. Every feeder must plan storage to suit his way of doing business. If he has a large farm on which he grows alfalfa, grain and other crops that make large quantities of roughage he must provide an extensive storage barn with appliances to get the stuff in and out again when needed for feeding.

Generally speaking the barn should be large and high. The capacity of a storage barn is increased by additional height at a very rapid ratio because all kinds of loose fodder packs very close in the bottom and lies very loose at the top. A deep bay may be filled to the peak with hay at haying time and settle sufficiently to hold a large quantity of sheaf wheat a few weeks later, but a shallow mow don't hold much at any time. It don't have the weight sufficient to pack it.

There will, of course, be a good solid floor over the car track and there will be chutes or openings to let the hay down directly into the car and there will be a ladder to let a man down into the car to tramp it full. The same horse fork that is used to put the fodder in will move the stuff from the other parts of the barn to this floor as it is needed.

Brick Pavement.

There is only one way to have a feeding lot clean and that is to pave it. There are different kinds of pavements more or less virtuous but the cheapest satisfactory bottom for a feeding yard is brick laid on a foundation of sand and cinders. The cinders help drainage and prevent the bricks heaving with the frost. It is easier to lay the bricks level and smooth if an inch or two of sharp sand is scattered over the top of the cinders. The sand holds the bricks in place and a little sand does not prevent the water from getting away.

A great deal depends on the foundation. The ground should be graded with the proper slope to the center gutter. It is not necessary to have an opening in the bricks, the cracks between the bricks are sufficient, but a line of tile should be carefully laid underneath deep enough to be out of the way of frost. Frost does not penetrate deep in a feeding yard under a brick pavement. During some winters the ground won't freeze. There is more or less litter scattered about that prevents hard freezing. Probably if the tile starts a foot below the brick at the shed end and deepens to two and one-half feet where it joins the trunk tile in the alley the drain will give no trouble.

Lay the tile first smoothly and evenly and cover the joints with pieces of broken tile, then fill in with coarse cinders using no earth over the tile. Tile in a mud-bottom barn yard seldom works satisfactorily because the tramping of the cattle packs the mud so that the water can't get through. A mud-bottom yard has never been drained and the chances are that such a yard never will be drained in a satisfactory manner.

Commence laying the brick in the center over the tile and work both ways to the fences. The herring bone style of laying brick gives the best satisfaction. No two brick tip alike when laid like this. Of course you want every brick to lay flat and level, but you don't always get just what blowing under and the ground floor of the shed slopes to the brick pavement. A liberal supply of straw for bedding is kept in the shed and this is carefully shaken up every day.

Feeders now-a-days appreciate the importance of making animals comfortable. It takes a good deal of feed to supply the heat dissipated by animals lying on the cold ground. Straw is cheaper than corn.

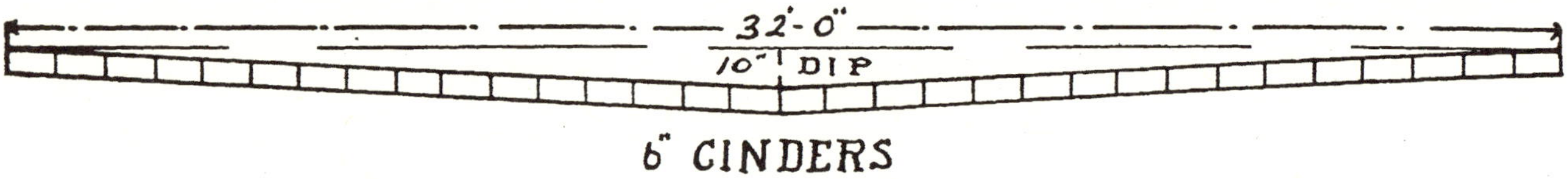

DETAIL OF BRICK PAVEMENT

you want. If good hard burned bricks are laid flat, herring bone style on a good foundation you will have more comfort and satisfaction than you ever had in a feeding lot before. If you have lots of money to use and don't care for expense then put in a cement pavement and build it just the same as sidewalks are built. You will then have a yard that will last a lifetime, but it won't be as dry as the brick because the water must all run to the end or center outlet on top of the pavement before it can get away.

The Shed.

A continuous shed is designed to run the whole length of the feeding plant without a break. The shed is twelve feet wide and eight feet high in front and six feet six inches high at the back. The shed faces the south and the front is left open to admit sunshine. The construction is light and cheap as shown in the detail drawing. There are no partitions except the fences between pens which run to the back of the shed, in fact the fence posts and shed posts are the same.

Two by six rafters fourteen feet long are used for the roof. These are covered with sheathing boards, dressed one side, and on this is stretched a good quality of felt roofing. The north side is banked with cinders to prevent the cold winds from

Beef cattle don't require much protection against the cold. Their thick winter hair and hides are sufficient if they are kept dry and well fed. Cattle will gain a little faster on the same amount of feed if kept warmly stabled, but they must have fresh air and the extra expense of individual attention when handling them in a stable more than eats up the additional profits from the extra gains made. A feeding rack well up above the ground along the back of the shed is a good thing at times in rainy weather; it induces the cattle to stay inside. It is better to put the feeding racks on the ground when you use them regularly every day, but ground space in the shed is limited and such racks will be used occasionally only For this reason it is not desirable to take up any more ground space than necessary for this purpose.

Corn Crib.

On the south side of the alley way is a corn crib six feet wide at the bottom, eight feet wide at the top, ten feet high above the foundation posts and as long as necessary. This corn crib is intended for storage purposes to hold corn enough to last all winter. There is a door in the end and doors along the alley side thirty-two feet apart, each door being opposite the door of a feeder crib. A temporary bridge reaches from one door to the other so the

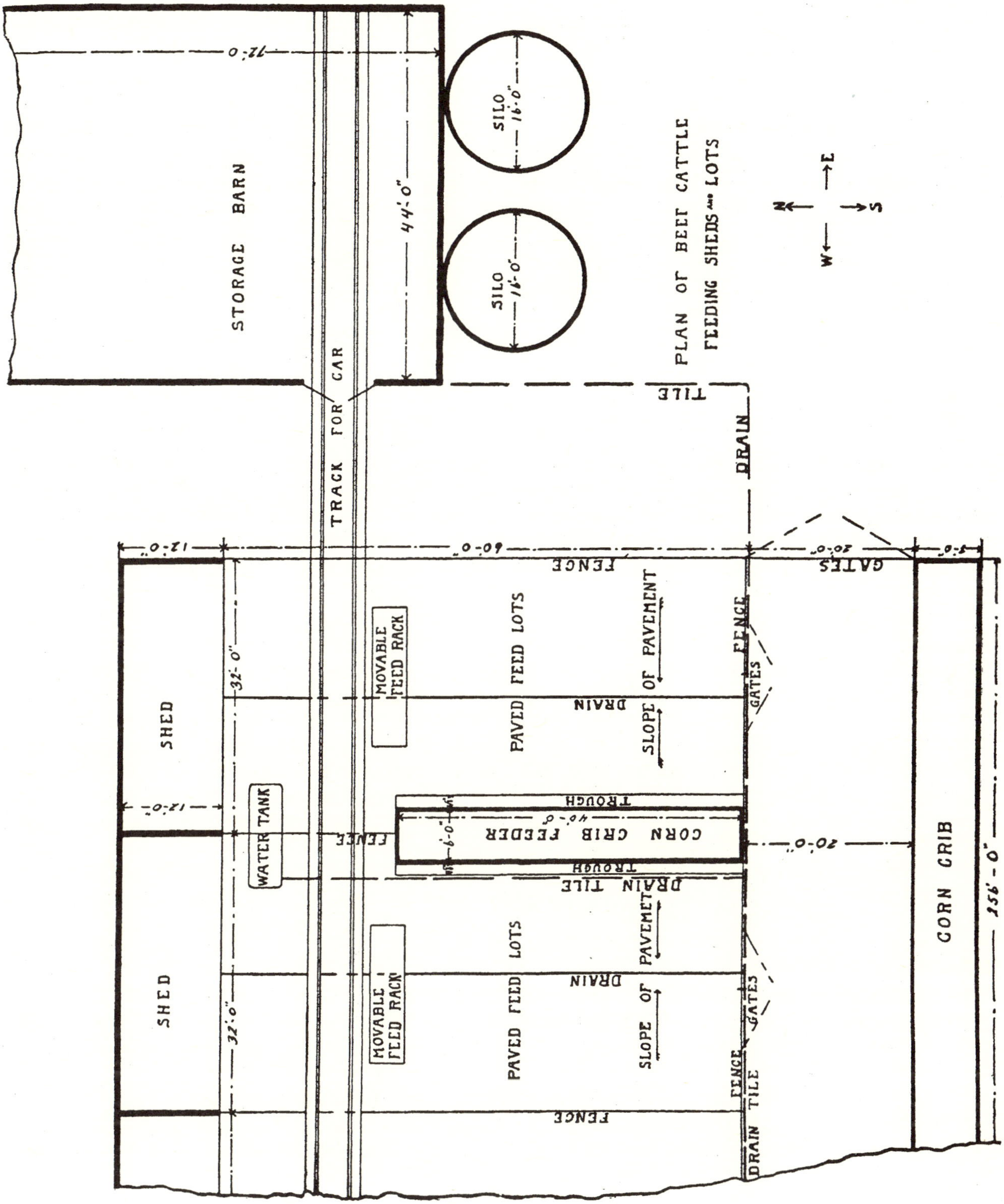
PLAN OF BEEF CATTLE
FEEDING SHEDS AND LOTS
STORAGE BARN
72'-0"
44'-0"
SILO 16'-0"
SILO 16'-0"
TRACK FOR CAR
DRAIN TILE
12'-0"
60'-0"
20'-0"
5'-0"
FENCE
GATES
SHED
32'-0"
MOVABLE FEED RACK
PAVED FEED LOTS
SLOPE OF PAVEMENT
DRAIN
WATER TANK
TROUGH
CORN CRIB FEEDER
40'-0"
6'-0"
CORN CRIB
256'-0"

carrying may be done with a wheelbarrow or car running on a track. As the bridge is intended to be moved from one feeder crib to the next a wheelbarrow would be handier than a car because it is lighter and may be easily moved.

Feeder Cribs.

Between each two pens is a feeder crib six feet wide at the bottom, eight feet wide at the top and eight feet high. These cribs are forty feet long extending back from the alley fence. This gives forty lineal feet of corn trough for each feeding yard. These feeding troughs are made by extending two by four floor cross joists two feet beyond the sills at each side. The floor in the crib is laid on top of these cross joists and the feeder boxes are made by boarding on the under side and across the ends. This makes the floor of the feeder trough about five inches lower than the floor of the crib which permits the corn to work out easily and in case of a driving storm the water does not run in from the feed troughs to wet the crib floor.

Some little experimenting is necessary to get the opening the right size. A smaller opening answers when the trough is lower than the corn floor. A narrow strip may be nailed in the opening at the top if it is found too large.

The roofs of these feeder cribs are made by using sixteen foot boards full length. The projection keeps the feeder troughs dry and provides a little shelter for the animals when feeding. For the comfort of the cattle it is a good plan to run eave troughs the whole length of these roofs. The water could be carried to the water tanks or the drain in the alley.

At corn harvest time these feeder cribs of course would be filled first with the earliest and best seasoned corn to feed first. The later and poorer quality of corn would be housed in the main storage crib.

It is not every feeder of beef cattle who approves of self feeder cribs, but if they don't like to have the animals help themselves the same cribs and the same troughs will be just as useful, so that the man who really loves to work may dig the corn out, load it in a basket and carry it around to the side of the crib and distribute it along the troughs. It will pay some men to do this, men who are built that way. Each man must work in his own harness.

Silos.

For some unaccountable reason beef men have entertained a prejudice against silos. But not every man who feeds cattle without their assistance objects to silos. In many cases they have more corn stalks than they can feed without trying to save the last vestige of the corn crop and they think the animals can cut the feed and mow it away cheaper than it can be done by machinery, but the fact remains that nearly one-half of the feeding value of the corn crop is in the stalks and leaves of the corn plant. If cut just at the right time, when the sap is all in the stalk, cut up fine and packed away in an airtight silo the stalks lose very little of their feeding value. They may be kept a year and the last silage from the bottom comes out as fresh and apparently as palatable as the first. Cattle will even leave pasture in the summer time to eat left over silage. If we ask the animals what they think of it their actions are strongly in the affirmative. We must study these things in detail to thoroughly understand our business.

Looking at the silo problem from the broadest side it certainly would pay to put some of the crop in silos. The stalks from eigth or ten acres will fill a sixteen by thirty-two foot silo so that most feeders would only have an opportunity to cut off one side of the corn crop and they would still have a large quantity of stalks to go to waste.

The silos in the plan are made of two by eight pine planks dressed both sides, the edges beveled and put together like a tub. They are hooped with three-quarter inch round iron hoops drawn up with nuts against the shoulders of cast iron plates as shown in the detail drawing on another page.

This feeding plant is designed to save

labor and to utilize feed to the best possible advantage. It would be difficult to build a large plant any cheaper and have it satisfactory. It would also be difficult to build, on any other plan, a thoroughly practical plant that could be extended indefinitely as the business grows without altering or rebuilding.

CHEAP CATTLE SHED—A123

Some kind of a cattle shed is necessary in connection with every feed lot. Plan (A123) shows a cattle shed ninety feet long and ten feet six inches wide. It is built of two by fours for framing, covered with boards twelve feet and sixteen feet long which cut to advantage without waste except at the ends.

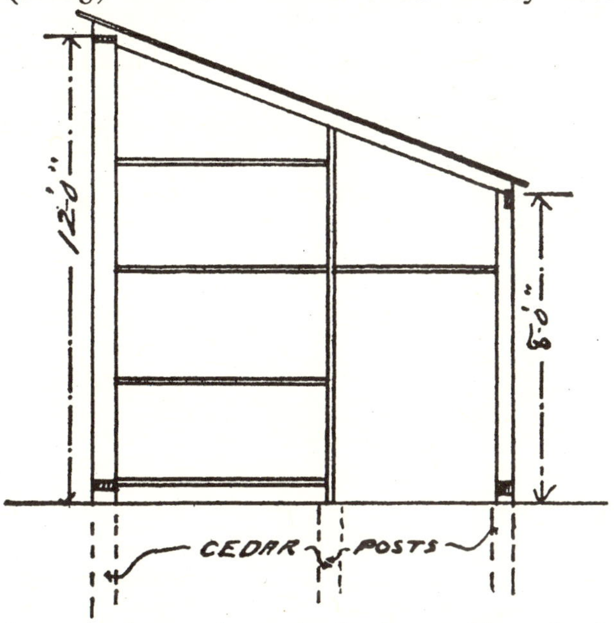

There is a low-down manger which runs the full length of the shed against the back wall. The front side of the manger is bedded in the ground which together with a little banking on the outside prevents the cold winds from blowing under. Some feeders fail to realize the importance of this precaution. The north wind seems much colder when it forces through a small opening. There is something about the bottom of a shed that seems to invite a current of air from the north, but this feed manger arrangement seems to get the better of it. Mangers should be low for another reason. For thousands of years cattle have been accustomed to feed from the ground. While in pastures they keep their heads down nearly all of the time, but for some unaccountable reason they are expected to hold their heads two or three feet high when being fed artificially.

The shed is supported by short cedar posts which are set well into the ground, the tops of them being cut almost even with the surface. The doors are made wide enough and high enough to get in easily with a manure spreader, and there are no posts or partitions in the way so that it is easy to clean out the manure.

SECTION AND PLAN OF CATTLE SHED

CATTLE SHED—A155

Sheds on three sides of a hollow square is an old style way of building feeding sheds. It is probably the best way now except that it is more difficult to economize labor with this construction than it is with a straight away proposition where you can run a railway and a feed truck the whole length of the shed. The hollow square proposition has the advantage of warmth because it is protected from the east, west and north winds. Yards like this are always built opening towards the south.

swing easily and fasten quickly help a good deal.

It is customary to drive around with a

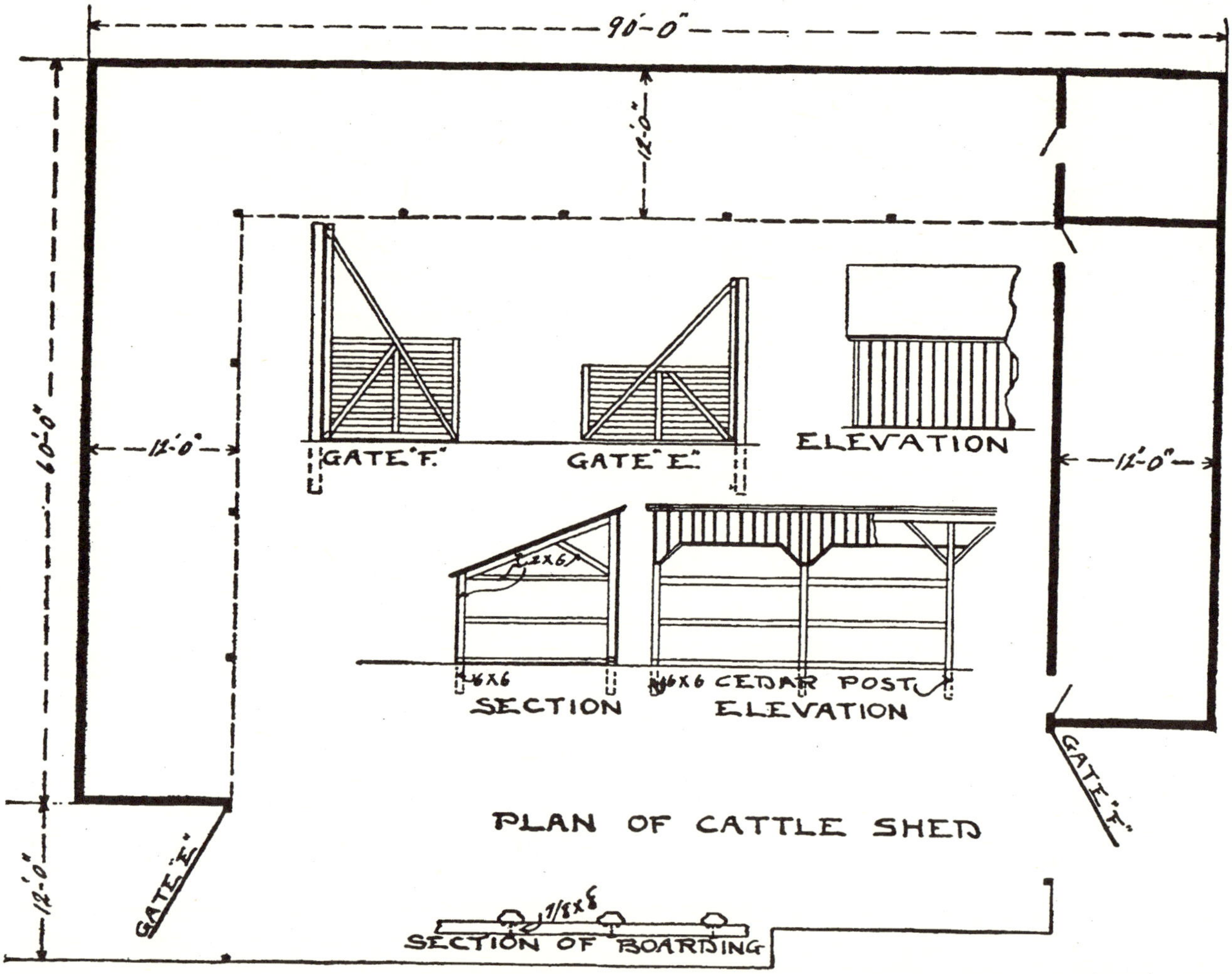

In this plan there are convenient gates to drive in when bringing roughage or other feed to the cattle. The gates to look well should be made right and left and they should have automatic devices to fasten them quickly. Animals confined in a yard in the winter time are crazy to get out. They learn how to slip up alongside of a wagon and crowd through the gate when the driver is engaged with his team. This is a source of annoyance that can hardly be avoided, but good gates that rack load of feed and dump a little in each feed rack as often as necessary. Sometimes a self-feeder for corn in the ear is placed in the middle of the yard and this helps a good deal in saving labor and the labor problem is worrying feeders more every year. There are feed carriers that may be hung from an overhead track to pass around through a shed like this, but usually the cars do not hold enough to effect much of a saving.

Poultry Houses

THE BEST POULTRY HOUSE—A219

THE cheapest poultry house may not be the best, but it has been demonstrated that it is not necessary to go to a great deal of expense to make a poultry house that is as good as the best. A few fundamental principles will cover the whole subject. A poultry house must be clean, airy and dry, so the location is very important.

If you ask any experienced poultry man what one thing has given him the most trouble he will tell you lice. If you chase down a failure in poultry raising and get at the true inwardness you will find lice at the bottom. Knowing this to start with one of the first requisites is to build a house that may be easily and thoroughly cleaned and kept clean. Have the word cleanliness stamped on the house from one end to the other.

To accomplish this every article of furniture in the house must be removable. There must not be a crack or crevice to harbor lice that you cannot fill with crude oil or some other disinfectant and you must have a handy cleanout where the litter may be removed and replaced with new clean material with as little work as possible.

Next to cleanliness comes ventilation which in a poultry house means admitting plenty of fresh air without the slightest suspicion of a draft, a problem that has caused more experimenting than any other one item in connection with poultry raising.

After years of experience the whole problem has been worked down to a very simple construction and the most popular house today is a low, cheap affair that may be built by any intelligent farm hand if he will simply read and follow instructions. Poultrymen differ in regard to the width of a house. Some want a house from twelve to sixteen feet wide so the sun can shine clear to the back of it; other poultrymen want a house from sixteen to twenty-four feet wide so they can house more poultry at practically the same expense. We are showing the general plan and giving the reason why without specifying any special width. It is understood of course that these houses may be any length as one pen is a duplicate of the other all the way through.

The front of the house is from eight to twelve feet high, according to the width, but three and one-half feet is high enough for the back of any house, and it is better low because the roosts are placed back here where it is warm in winter and you want a low ceiling to confine the warm air close to the fowls. You can secure ventilation by having a warm roost. The body heat of the fowls will warm the air and we all know that heated air is lighter than cold air and for this reason it will follow the slant of the roof upward and cold air will come in from the front to gradually take its place. In this way you can get a constant change of air around the fowls without a draft.

We have windows in the high front side to admit sunlight and air. Some of the windows are covered with cotton, while some of them are fitted with glazed sash. The extent of the cotton surface depends upon the size of the house, the number of fowls kept in a pen and the climate. In the far north you cannot have a great deal of cotton surface without making the house too cold, but right here is a point to remember. Fowls are warmly clothed with feathers and they will stand the cold a great deal better than bad air provided that the cold is dry, and it is understood that the poultry house is dry otherwise it is no good.

Another important feature is the kind of cotton. Cotton does not mean canvas or thick heavy ducking, it means something that will let the air through freely. Some poultrymen say the thinner the better, but probably a cheap grade of cotton such as you buy retail for about four cents a yard will answer the purpose best. And

FRONT ELEVATION

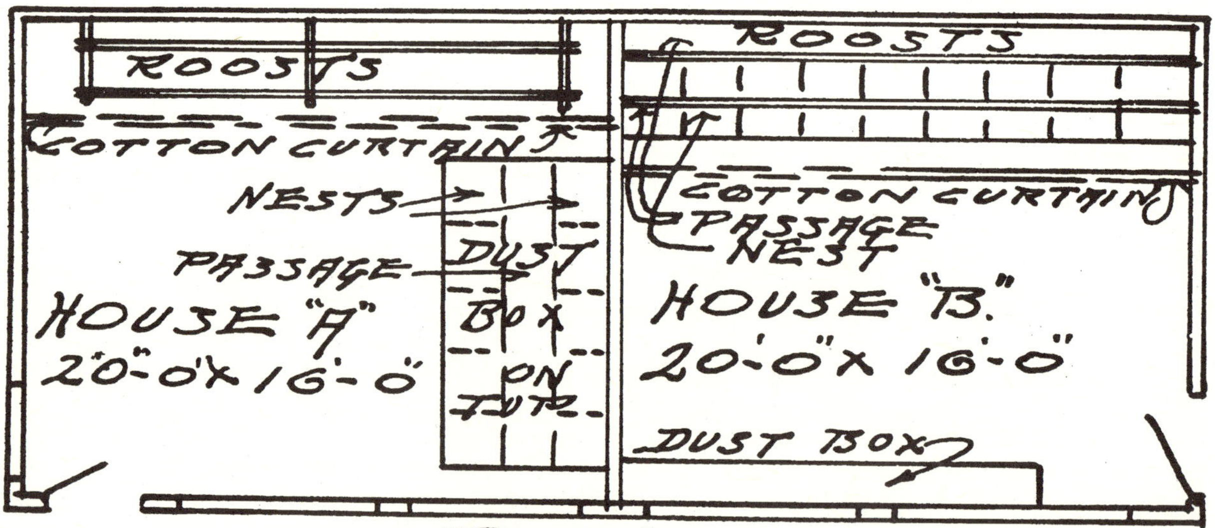

PLAN

you want to tack the cotton on loose, leave it baggy so it may flap in the wind because every flap jars the dust loose. If the cotton is drawn tight dust will settle and clog the little openings.

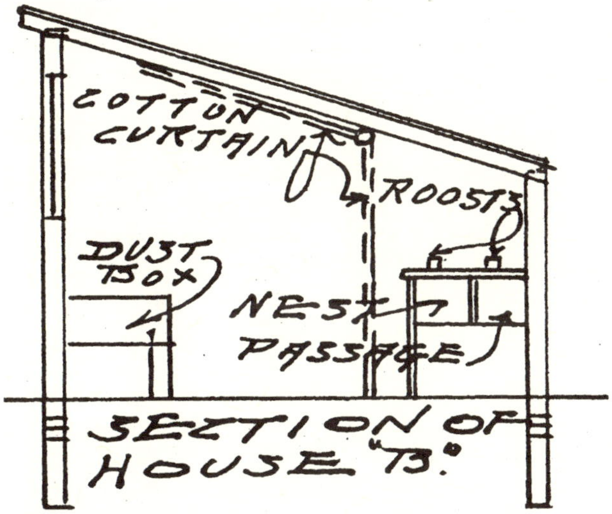

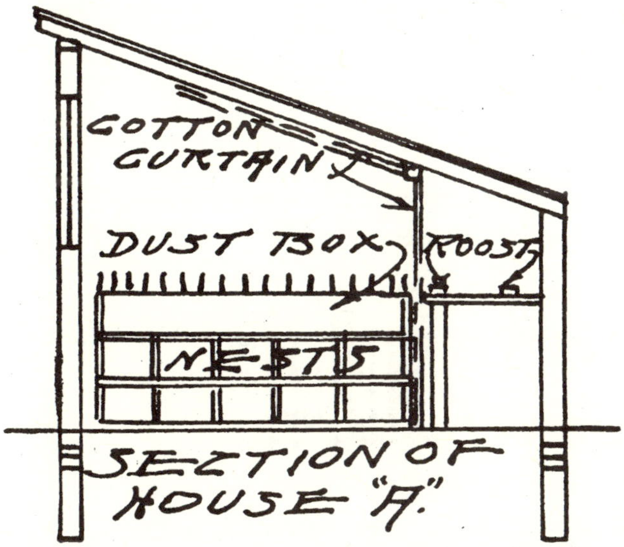

Some poultrymen cover all the openings with cotton and have glass sashes inside that they may slide shut at night or during extremely bad weather. When the cotton ventilators are made right and kept right the air in the poultry house will be good.

In the north make no openings into the poultry house except in front, unless you have front yards to interfere. In that case make the doors in the ends right against the front corner.

For material use matched stuff with building paper either one or two thicknesses according to the degree of cold you are obliged to guard against. A good house is made with a light frame work of two by fours covered with seven-eighth matched boarding dressed smooth on one side and the smooth surface turned in. The two by fours are dressed on all sides to make them smooth. Avoid every possible roughness because it will hold dust. Dust will harbor microbes as well as lice and mites, which leads us back to one main branch of our text, cleanliness.

Have no eave projection at the back. Cover the matched boarding with tar paper commencing at the back at the ground and work up to the high part of the roof in front. Make good joints nailed down into fresh tar. Put the tar roofing paper on in such a way that it will hold standing water, then you know that neither water nor air can get through. Cover the ends of the building in the same way.

Dropping boards are placed against the back of the house fitted close against cleats to prevent draughts and hung with hooks so they will lift up out of the way to clean the floor.

Six inches above the droppings boards are loose roosts that may be easily lifted off and carried outdoors for cleaning. The oftener they are carried out and given a heating up in the sun the better.

Some poultrymen prefer what they call nesting rooms. These may be easily made by placing two rows of nest boxes together with a passage between so built that they may be easily moved about or carried out doors. Room A shows such an arrangement. There is a large dusting box on top with wooden pins projecting to prevent roosting on the edge of it. In room B the dust box is built on legs so it stands up eighteen inches to two feet above the floor. This leaves the ground space clear for litter for scratching and keeps the contents of the boxes free of straw and trash so the hens can dust themselves better. Then a box elevated a little catches the sunlight through a window to better advantage.

Hens love to dust themselves in the sun in winter time. If the boxes are large enough to hold three or four hens at once it is all the better because it fosters a sort of hen sociability. If you watch them at such times you will notice that they apparently talk together and enjoy a little hen gossip among themselves. It is only by studying the habits of poultry that we can get at the proper management. Poultry management to be successful depends on little things and there are a great many of them.

This house may be built directly on the ground, or it may have a cement or stone wall under it. The floor inside may be earth or cement, but never of wood. Earth is the best, cement is rat proof, but no matter what the floor is you must have the sills so imbedded that there will be no draft. This may be done in a dozen different ways. Perhaps the simplest is to nail a board on the sill that reaches down into the ground eight or ten inches. There is no hard and fast rule for the way in which it is done so long as you accomplish the purpose, that of keeping the air from blowing through. You want air but you don't want to admit it under the sills or through cracks in any part of the building.

For cleaning the droppings boards you need a half barrel mounted on wheels and it should be used every day. One great value of such a convenience is that it will be used when an awkward affair would be neglected as much as possible. A cart like this costs very little and is worth a great deal. For the same reason we want some kind of an absorbent to sprinkle over the roosting boards. Roosts may be kept clean if you go about it in the right way. It is well worth the effort.

PRACTICAL POULTRY HOUSE—A168

A single section of a two-pen poultry house fourteen by twenty-four feet is given in this plan. The house of course may be any length by adding any number of twenty-four foot sections. It is placed so that the windows look to the south to gather all the sunlight possible.

PERSPECTIVE VIEW

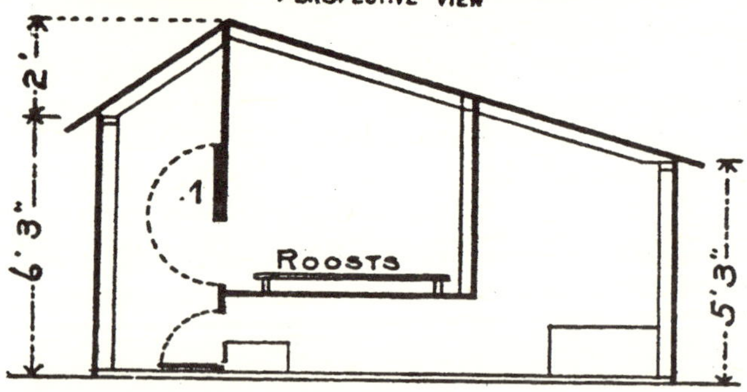

SECTION.

A passageway on the north side, where the roof is high to make head room, is partitioned off and the work of feeding is done along this passage. A door lifts up in front of the roosts from this passageway to facilitate cleaning. It is not necessary to enter the scratching room very often because most of the attention may be given from the alleyway. With the exception of the space occupied by the dust boxes the whole floor, except this passageway, is given over to scratching purposes as the roosts and dropping boards are elevated so the chickens can work under

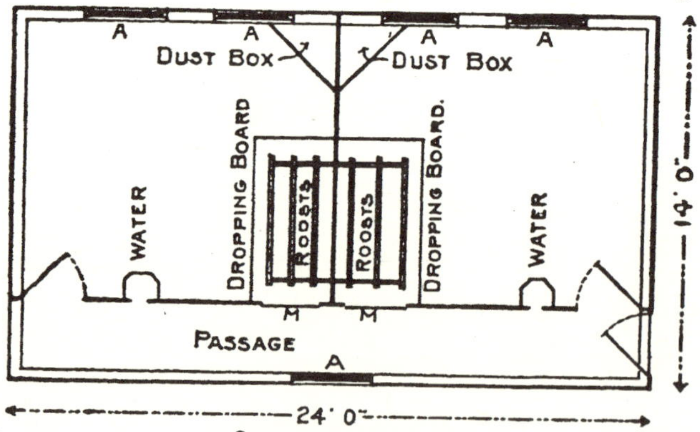

GROUND PLAN.

them. A section of this house will accommodate from twenty to thirty birds according to size.

Poultry men argue by the hour about the necessity of an alleyway. There are many different opinions. Some think an alleyway is worth all the room it takes up just to prevent annoying the fowls, when feeding by going in and out from amongst them. Other poultry men think that chickens ought to be tame enough to pay very little attention to the feeder when he goes about his work, but it is generally noticeable that a hen makes quite a fuss when she thinks she is about to be cornered. This applies to hens that are ordinarily tame, as well as those that are ordinarily wild.

HEXAGONAL POULTRY HOUSE—174

The house shown is in the shape of a hexagon and makes a very handosme and convenient house, and is just the thing for the city lot where space is limited. The ground or floor plan will show you the interior arrangement. The size of this house is ten feet six inches, and each of the sides is six feet three-quarters inch in length. The corner posts are six feet long and the center of the house nine feet from floor to peak of roof. The house should be built with one window facing directly south and the other facing southeast, thus allowing an abundance of sunlight to enter the building in the morning, when it is most needed.

ELEVATION.

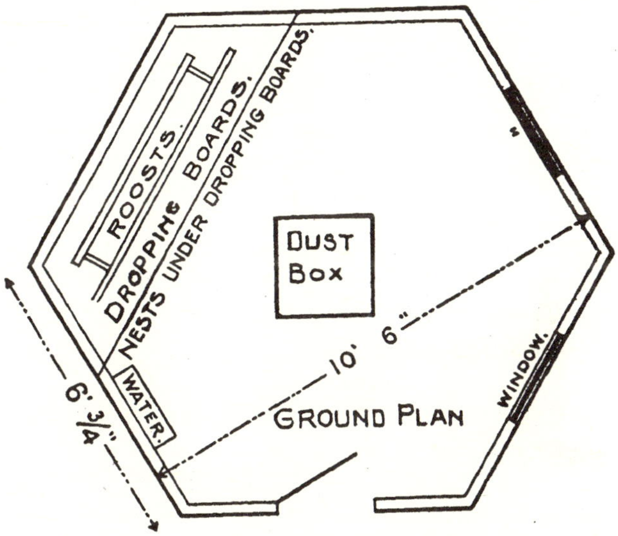

In nearly all the plans given it is designed that the ground floor shall be of earth, which is, in most cases, the most satisfactory floor material, and should be used whenever practicable. Cement floors are also good, however; where they are used the poultry house will generally present a more attractive appearance and can be kept cleaner, with less labor, than a house having earth or wooden floors. Wooden floors should not be used if they can be avoided.

OPEN FRONT POULTRY HOUSE—A110

The modification of the popular open front poultry house is given in this plan. It is suitable for two lots of hens of forty or fifty each, according to the size of the breed. The house is forty feet long and ten feet wide, divided into two compartments. Each compartment has a warm room and scratching shed which is open to the south. This makes each room ten feet square with a roof eight feet high in front and four feet at the back. No room is taken up in hallways or passageways but the doors entering the warm rooms open from the scratching sheds.

Very light material is used in the construction of this house. Sills are four by six inches, and two by fours are used for rafters. Common lumber is used for boarding, which is covered on the outside with building paper and the building paper is covered with thin matched sheathing. For the roof common sheathing boards are laid close together and covered with

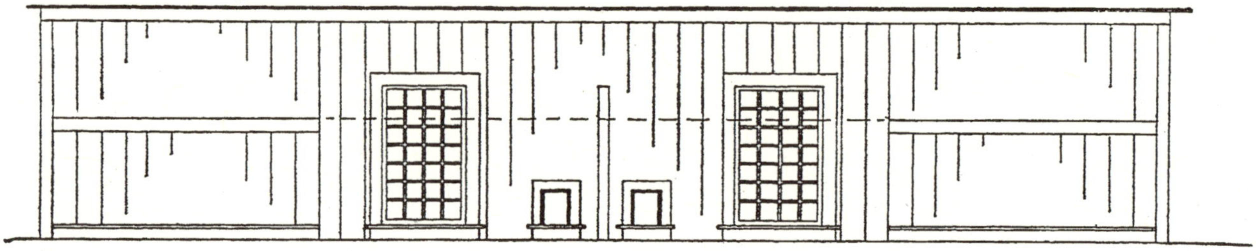

tarred paper and the paper covered with shingles. This makes a warm roof which is very essential to a poultry house.

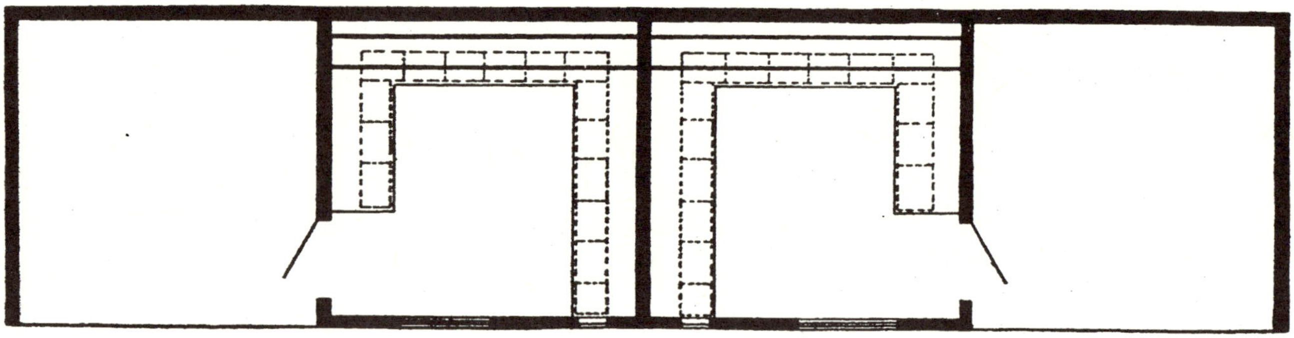

Each of the closed pens has a window that reaches down to the sill. This window is wide enough and high enough to let in a great deal of sunshine, and this is what the chickens need in winter. All inside surfaces are dressed to prevent lodgement of dust and hiding places for vermin. The whole bottom of the building is filled in several inches deep with grout mortar. In the warm rooms the floor joists are embedded in the soft mortar and a matched floor laid on. A floor like this is dry and easily cleaned and it is impossible for rats to work their way up through it. There is no wooden floor in the scratching sheds. The grout filling is supposed to be covered with straw a foot or so in depth. The hens will work in this straw even in the coldest days, but of course it is a good plan to have a liberal supply of straw in the warm room for amusement night and morning.

For nest boxes the arrangement given in this plan is very satisfactory. It shows a roosting platform with a row of nests underneath. For leghorns or similar fowls twelve inches square and seven or eight inches high is large enough for the nest boxes, but for brahmas or cochins two or three inches larger each way are much better. To facilitate cleaning the dropping board and nest boxes lift off from the lower platform. The lower platform is hinged and may be dropped down or unhooked and the whole thing carried out-

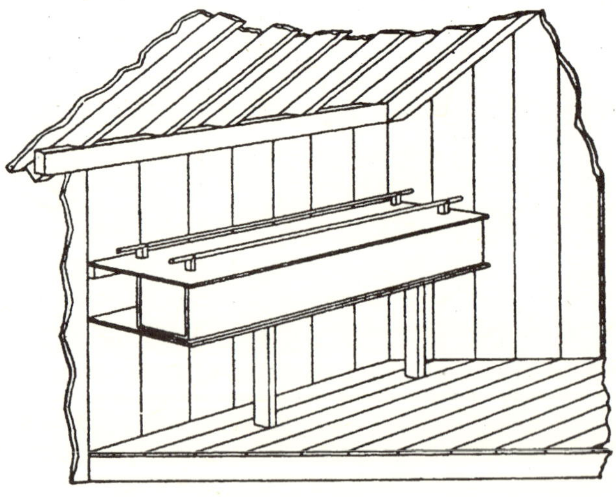

doors. It is very important to have roosting poles, dropping board and nest boxes loose. A great deal of trouble has come from vermin getting into these places without having facilities to eradicate them easily.

Hens seldom form the egg eating habit if the nests are dark. This is why the boxes open from the back under cover. The dropping board is not fastened to the nest boxes in any way. When gathering the eggs it may be lifted easily.

SMALL POULTRY HOUSE—A153

A little two story poultry house that looks like a plaything is shown in plan (A153), but this house is all right as far as it goes. It is especially valuable for a boy who would like to start in the poultry business but cannot afford a more expensive house. This little house is four feet wide and twelve feet long with a scratching shed the full size on the ground under the floor. This space underneath is two feet high and the windows should extend well across the front side.

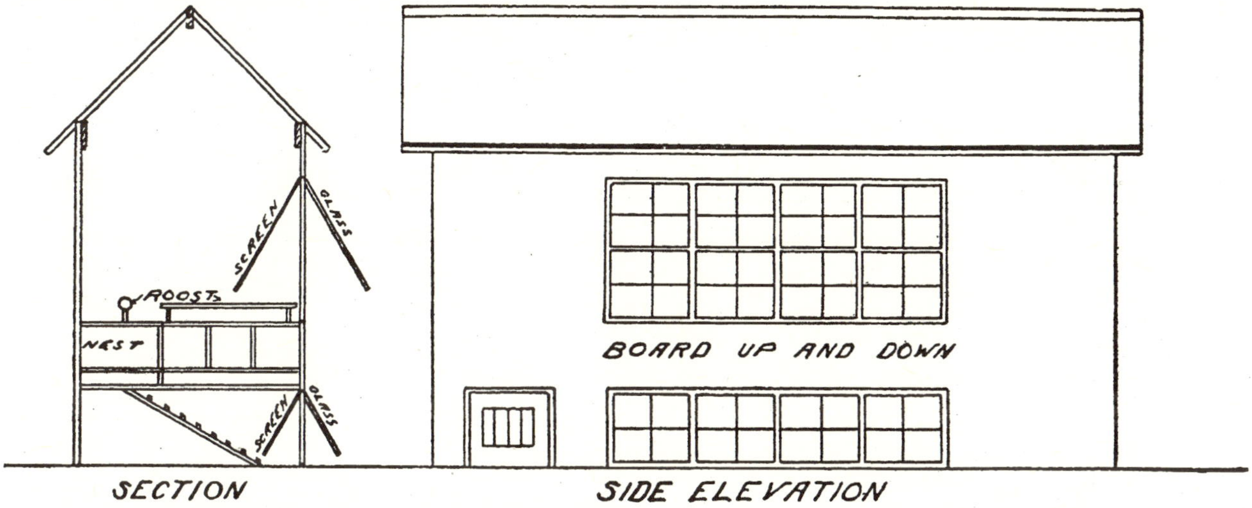

A runway for the chickens to get up and down the stairs is made by sawing off one wide floor board and hinging it in such a way as to let one end drop to the ground. When this is raised up it fits the opening in the floor and it should be fitted with jambs to keep the cold from coming through the cracks.

The nest boxes and roosts are loose so they may be moved about for cleaning or taken out at any time and put back as needed. It is not intended that any one will find it necessary to go inside this little house. The work is all done through the windows. The inner screens may be rigged with cords and pulleys to hold them up and the outside windows may be held up by braces from the building. To gather the eggs, clean out the house, or for feeding, one of the windows is raised and the screen pulled back with a cord. The chickens may be driven down stairs or upstairs during the operation. The screens may be of wire or cotton, or both. Cotton is the best because the window can be then left open and the chickens will get plenty of fresh air without a draught. On farms where considerable poultry is kept one of these little houses would be found useful occasionally to keep some breed separate.

DUCK HOUSE—A98

A house designed for the housing of thoroughbred ducks is given in plan (A98). It is built up from the ground on cedar posts set on blocks to prevent settling. The idea is to have a damp proof house with the best possible ventilation.

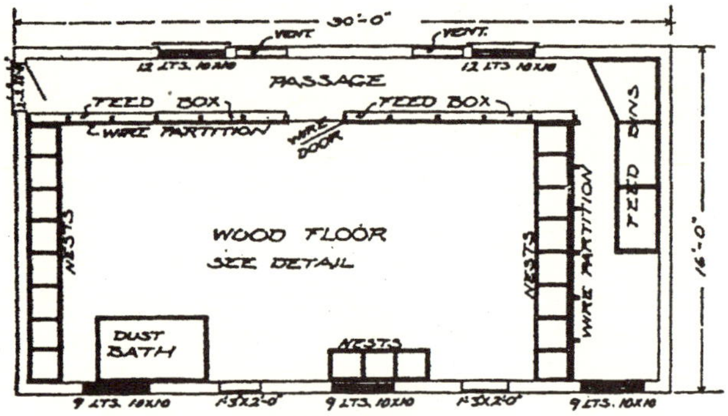

FLOOR PLAN

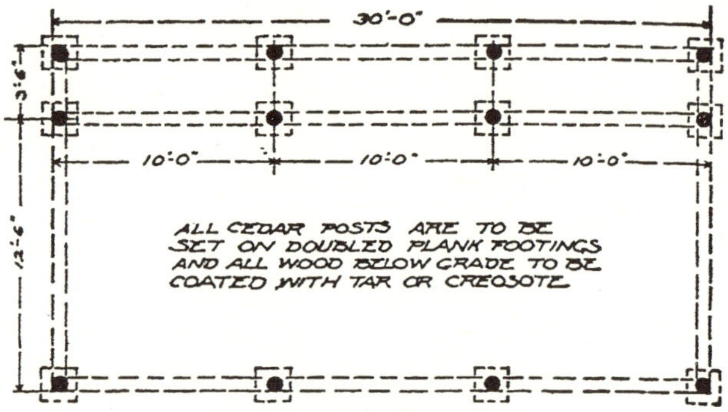

FOUNDATION PLAN

NORTH ELEVATION

SOUTH ELEVATION

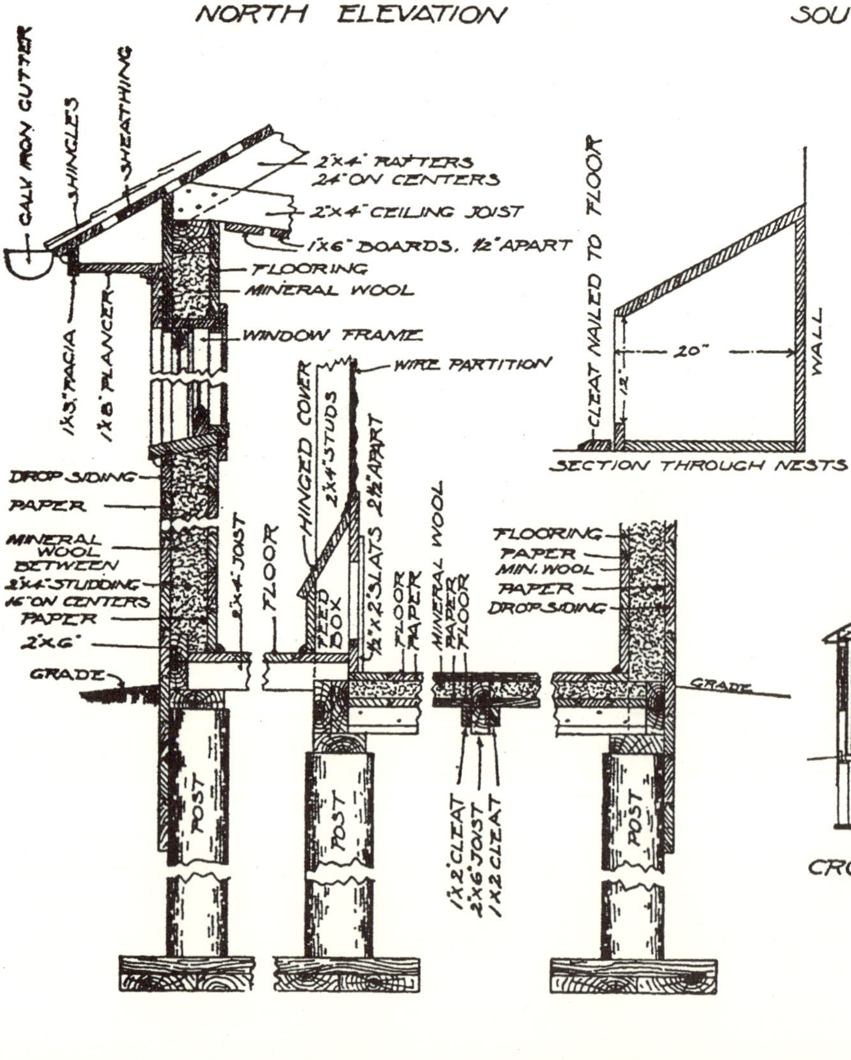

SECTION THROUGH NESTS

SECTION THROUGH DUST BOX

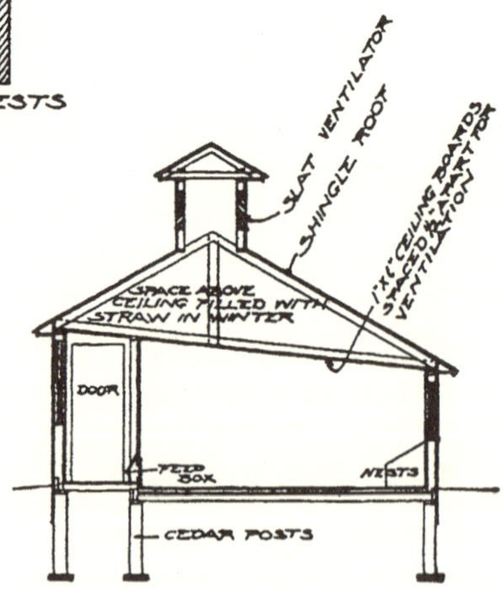

CROSS SECTION

The building is sixteen by thirty feet and contains one general room with a passage, which is also a storeroom for feed, along one side and across one end. All the principal construction details are fully shown in the detail drawings.

The house is built principally of two by fours as it is not very large and heavier timber is not necessary. The especial features are the filling of mineral wool in the partitions for warmth and a slatted ceiling with straw overhead for ventilation without drafts and without letting in an unnecessary amount of cold air.

This style of a building is somewhat expensive but it is very satisfactory when finished. It is usually considered that any kind of an old shed will do for ducks. In most cases any kind of an old shed is made to answer the purpose, but there is money in the better breeds and to get results it is necessary to keep even ducks with some idea of comfort. Some of the improved varieties bring fancy prices for eggs and young breeding stock, but like other thoroughbred animals fancy ducks need a little more attention than little old scrubs that most of us are accustomed to.

DOUBLE POULTRY HOUSE—A154

A small double poultry house is shown in plan (A154). It is twenty-four feet long and sixteen feet wide, giving a space of sixteen by twelve feet to each compartment. It is very simple and it is also cheap and durable. It may be built of matched stuff with the smooth side turned in, or it may be constructed of rough lumber. Of course matched stuff is very much the best as it leaves no harbor for vermin and no lodgment for dust. In either case the building is covered outside with tarred paper. The paper is started, in strips, from the eaves in front, carried over the peak and clear down to the ground at the back.

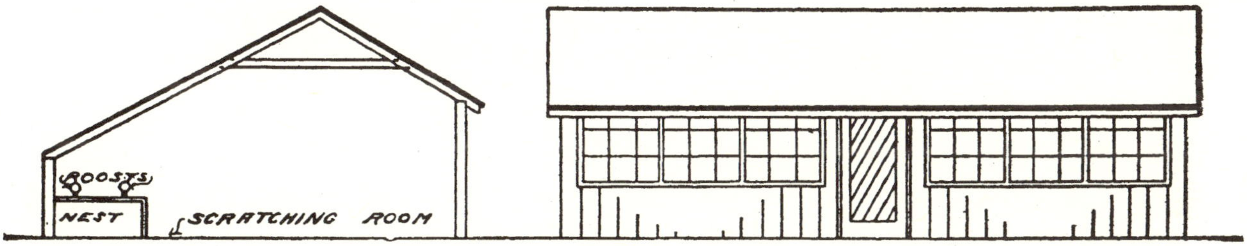

POULTRY HOUSE

Inside, the house is practically all one room, but a roost curtain may be hung with a roller to pull down at night or the cotton may be tacked on a hinged frame to let down at night, also one or more of the windows may be left open and the spaces covered with cotton.

Against the back wall is the droppings board with the roosts above it and the nest boxes underneath. All this furnishing is made removable so far as possible for easy cleaning. The apron board in front of the nest boxes lifts out in sections.

MODEL CHICKEN HOUSE—A173

This building is 68 feet long and 16 feet wide, built on a post foundation, which is enclosed with planking covered with galvanized wire cloth to a depth of about two feet below the ground, to check the tunneling of rats, etc.

Almost every lover of poultry has his own ideas as to how the model chicken house should be arranged and constructed, and every chicken house that is not thus constructed may meet with his severe criticism. We will, therefore, not lay stress on any one particular feature of this building but will say that several different ideas

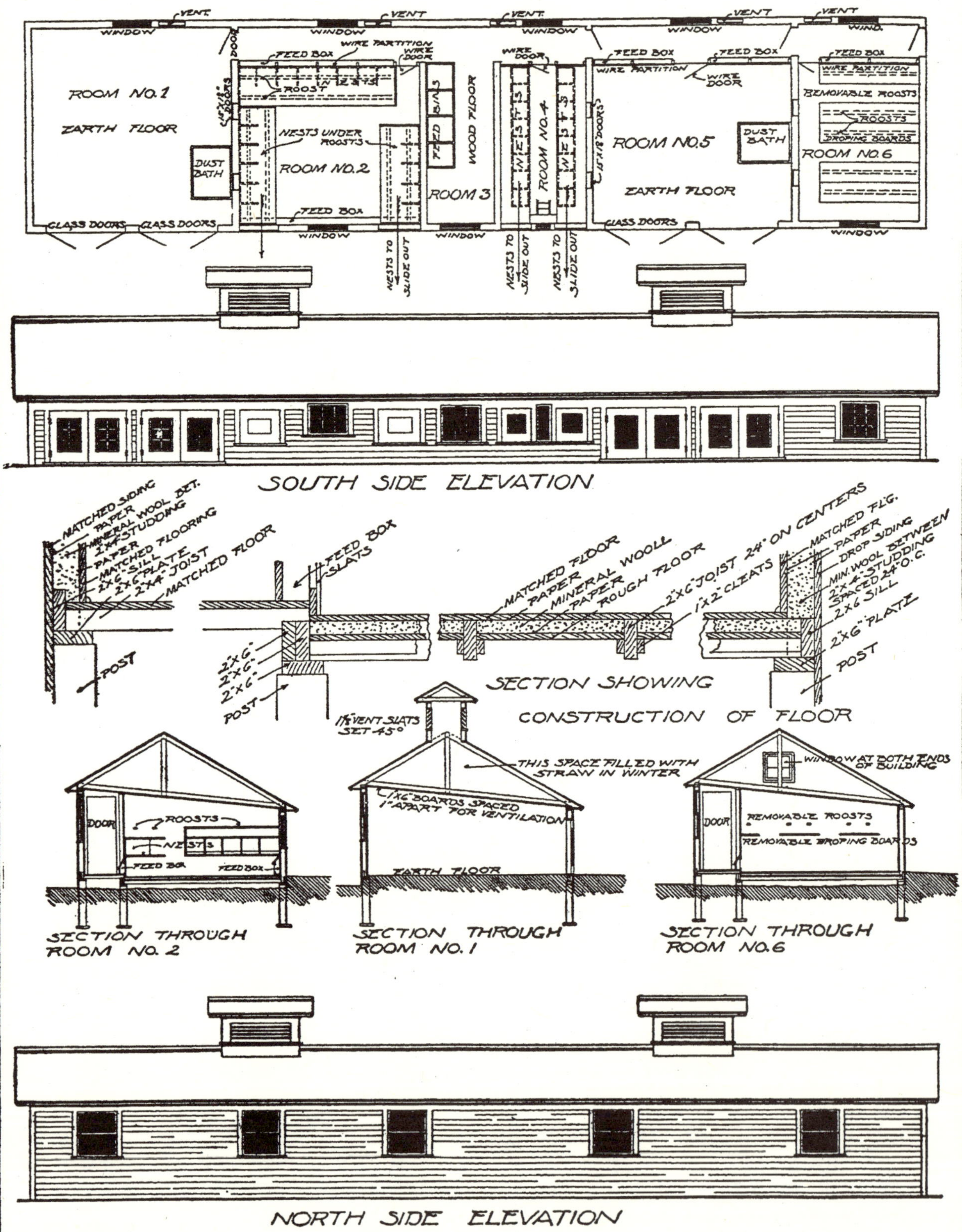

DESIGN OF CHICKEN HOUSE

have been used which may be explained as follows:

Rooms Nos. 1 and 2 (see floor plan) are used together; room 1 being the scratching room which is used in stormy and winter weather for exercise, and room No. 2 is the feed, nest and roost room.

The roosts are placed above the nests which have a cover, or roof, pitched so the chickens cannot roost on the nest, but are compelled to get on the roost above. The nests are open in front, having a passage for the chickens, running the full length of each section. The nest sections are removable through doors opposite each section, so they can be easily cleaned and aired; they set on a rack which elevates them about twenty inches above the floor, so the chickens can walk below them where the feed troughs are located, as shown in the section through room No. 2.

Room No. 3 is a feed room, 5 feet wide, which contains feed bins for grain, meal, etc. To the right (east) of this feed room are rooms 4, 5 and 6. In this scheme, the nest room, 4, is separated from the roost room, 6, one being to the west and the other to the east of the scratching rooms. This may have several advantages over the idea of room 1 and 2 where the chickens roost and lay in the same room, but it also has some disadvantages, one of which is that a larger building is required for the same number of fowls.

The nests of room 4 are so constructed that each nest can be taken out separately, or each entire section can be taken out through doors the same as in room 2. In place of the chicken being in view while on the nest, in room 4 the opening of the nests face the wall, having a dark passage for the chickens. By being out of view they are not frightened while the eggs are being gathered, which is done through a small round hand-hole through the back of the nest. This is covered by a small wooden shutter loosely screwed on over the hand-hole so it will always hang closed. Feed boxes similar to those in room 2, are located along the hallway.

Rooms 1 and 5 have earth floors and boxes filled with dust, for dust baths. All other floors are constructed double, with two inches of mineral wool between them for warmth, as shown in the section. All side walls of the building have heavy building paper both inside and outside of the studding, and the space between is also filled with mineral wool.

The space between the ceiling and roof is filled with straw during the winter months, and the ceiling boards are spaced half an inch apart to allow a free circulation of air through the ceiling and straw. This is brought about by having windows at each end of the building, which are controlled by cords. All windows on the north have storm sash for winter. Ventilation shafts are built in the north wall, with side shutters for admitting fresh air and exhausting foul air in winter, when all windows are kept closed.

SCRATCHING SHED POULTRY HOUSE—A151

A poultry house with an open scratching shed is shown in plan (A151). The house is thirty-four feet long by twelve in width. Poultry men differ about the width of a house constructed in this manner. Some prefer twelve feet because it is easier to get the sunlight clear to the back, as these houses should always front the south. On the other hand men with considerable experience prefer houses sixteen or even twenty feet in width because they can house more fowls for practically the same amount of money.

There are many ways of building an open scratching shed and poultry house, but this plan seems to contain about everything that is necessary. The door opening into the hen-house is just a frame covered with cotton which admits both light and air to the roosts and nest boxes. The outside wire netting may be covered with cotton or not according to the climate and the ideas of the owner.

The roofing is tarred paper and it starts at the highest point in front, turns over the upper corner at the back and goes clear down to the ground. This makes a thoroughly wind proof and damp proof house.

It is a peculiar thing about the damp-

SECTION

ELEVATION

ness in poultry houses. It is a comparatively simple question that has bothered poultry men more than anything else.

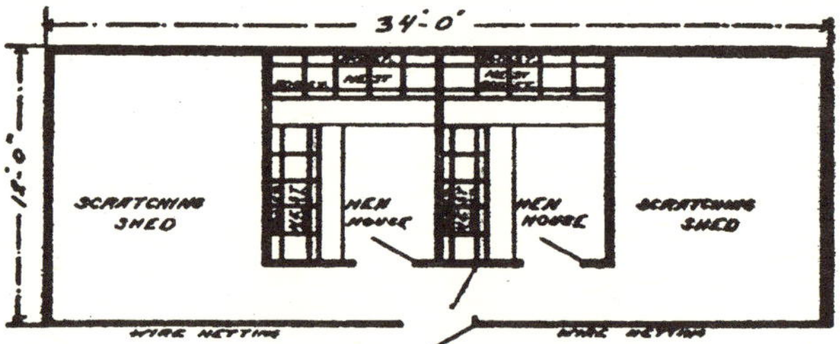

FLOOR PLAN OF CHICKEN HOUSE

Why a poultry house should gather dampness and have white frost on the inside when all the stables on the farm are comparatively dry has bothered more men than anything else in the poultry line. It is easier to build a satisfactory stable for any other domestic animal than it is for chickens unless we are satisfied with what is commonly termed a curtain front house. The phrase curtain front simply means that some of the openings are covered with thin cotton instead of glass. It seems to have solved the problem of how to make a chicken house light, airy and dry, but not all curtain front houses work alike. A great deal depends on the head room. A few hens have not body warmth enough to heat a great deal of space. You cannot have good ventilation without heat. The solution seems to be to build a comparatively small house with a low roof. Some poultry men build their curtain front houses as low as two feet at the back and only about six or seven feet high in front.

AN A-SHAPED POULTRY HOUSE—A152

An A shaped poultry house is given in plan (A152). This is the cheapest way to build a poultry house. You don't have to

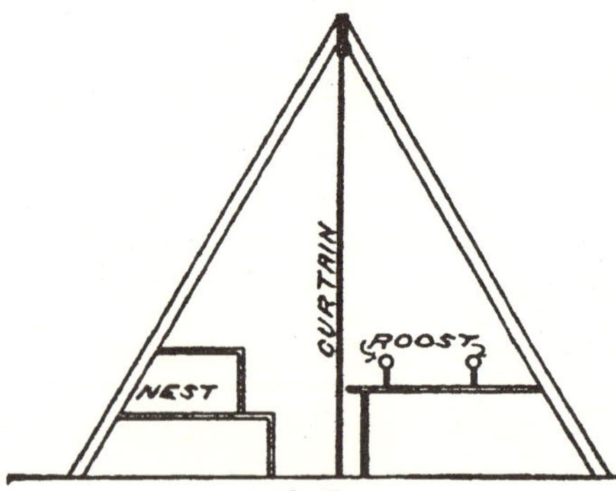

SECTION

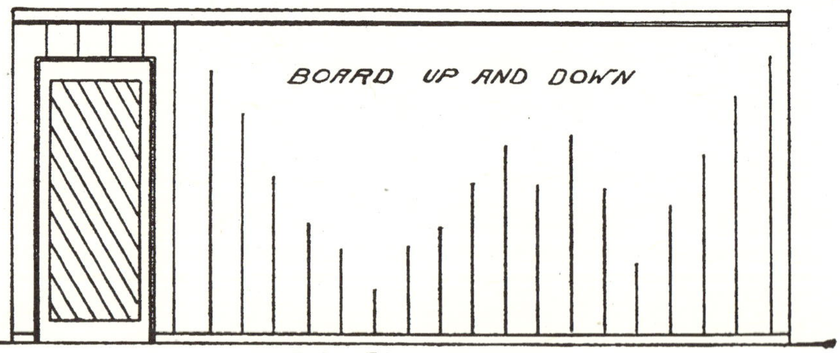

SIDE ELEVATION

build a roof or if you build a roof you don't have to build sides. You can do either way you choose.

It is divided lengthwise with a curtain partition. This curtain is in four foot sections and it rolls up on heavy window shade rollers, so that it may be pulled down cold nights to make a warm roosting place. The material of the curtain is cheap cotton costing three or four cents per yard. The sections are divided by two by four posts reaching from the floor to the ridge pole and made flush on the curtain side. You attach the roller to the ridge pole so the curtain rolls up on the inside of the roller which brings it close to the woodwork.

The house shown in the plan is eight feet wide and sixteen feet long. One end of this building is supposed to front the south. There is a small door in this end for the chickens to go in and out and the window is as big as possible. The entrance door is at the side and it should be near the south end. It is bad plan to have doors, windows or any openings in the north end or north side of a poultry house.

SMALL CHICKEN HOUSE—A119

A very neat little chicken house is shown in plan (A119). In size it is only 7x16 feet but it makes comfortable quarters for 15 or 20 hens. It is set on posts a foot or two from the ground to be out of the way of rats.

The dropping board, is loose and may be easily taken out through the door for cleaning. The roosts also are loose and may be removed easily.

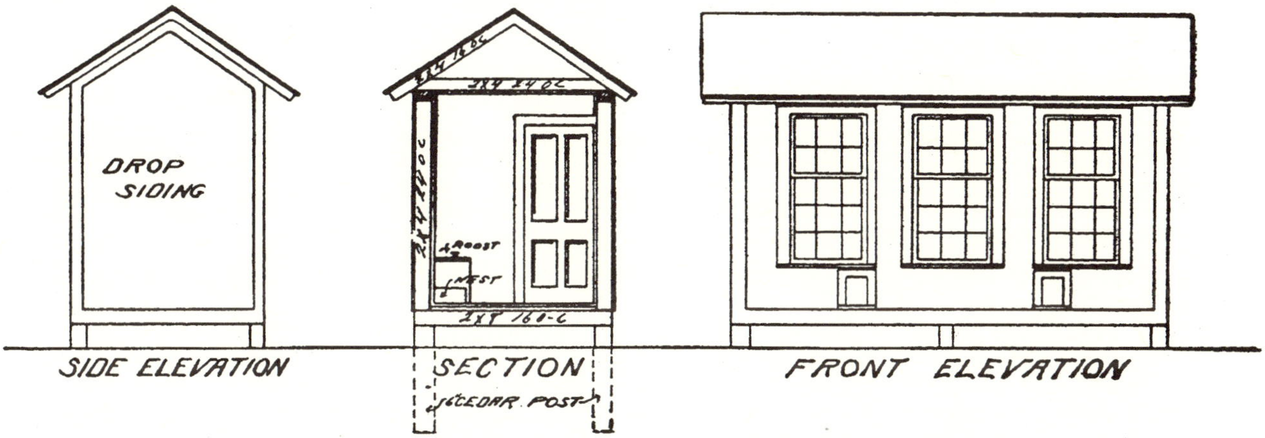

The floor is made warm by having it double boarded with a thickness of building paper between. The large windows of course face the south and the dust boxes are placed immediately in front of them because that is the way biddy likes to take a dust bath. She wants it directly in the sunlight if possible.

It is not necessary or desirable to go into a little house like this very often. It is so small that the presence of an attendant frightens the hens and causes a disagreeable commotion. By proper management, however, they can usually be let out into the yard when the presence of an attendant in the house becomes necessary. The roosts are placed over the nest boxes and the entrance to the nest boxes is in the rear. The nest box cover, which also is the dropping board, is loose and may be easily taken out through the door for cleaning. The roosts also are loose and may be removed easily.

This is just the kind of a house to start a boy in the poultry business. Boys take more interest in a small poultry house than they do in a house full size.

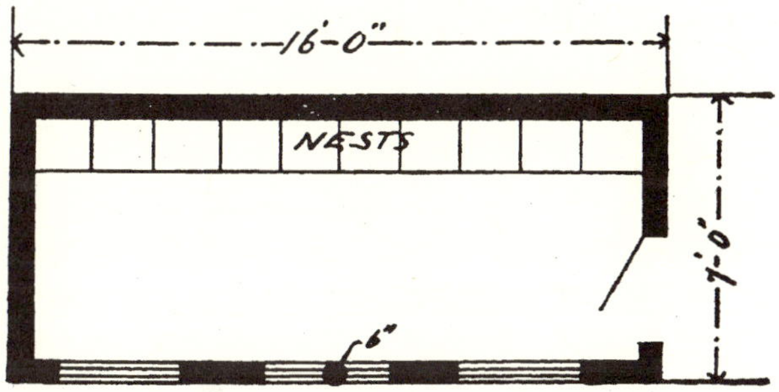

A little house like this is helped out very much by having a good yard in which considerable green stuff may be grown for the fowls to pick at. By planting a little grain and a variety of vegetables, the poultry will pick up a good deal of feed and the fowls will be more healthy because of it.

ELEVATED CHICKEN HOUSE—A165

This plan elevates the poultry house about fourteen inches above the ground for the purpose of preventing rats from making nests under the floor. It is high enough so that cats and dogs can have free access underneath and this space also offers a shady protection for fowls in the summer time. At the approach of cold weather in the fall this space is boarded up and manure is banked against the boarding to keep out the cold. If horse manure is used considerable heat may be generated.

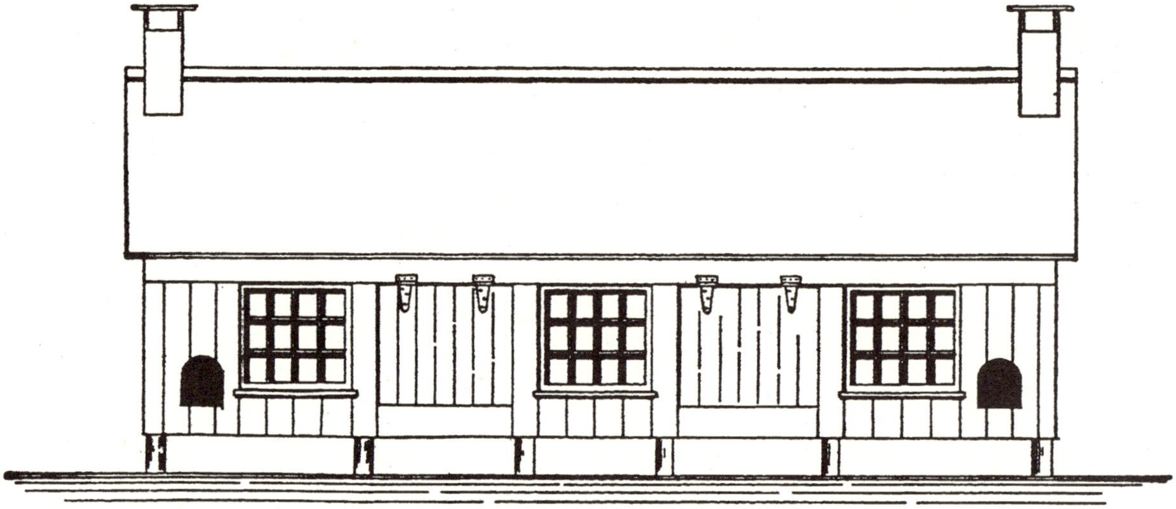

As the building is not very heavy the sills are made of two pieces of two by six, one laid flat on the supporting cedar posts and the other turned edgewise as shown in the drawing.

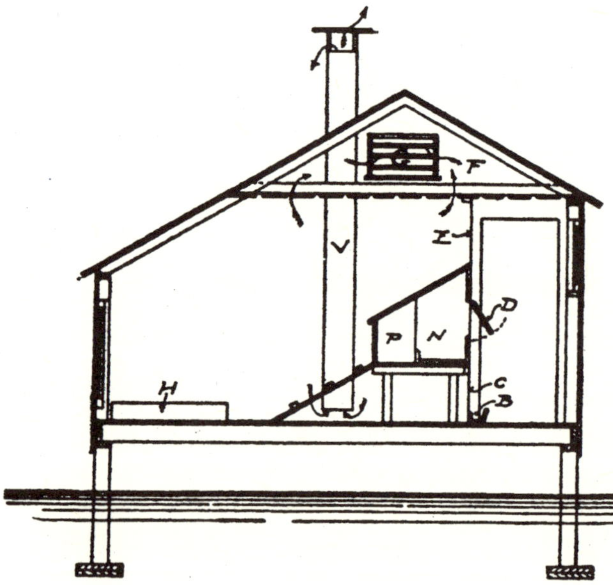

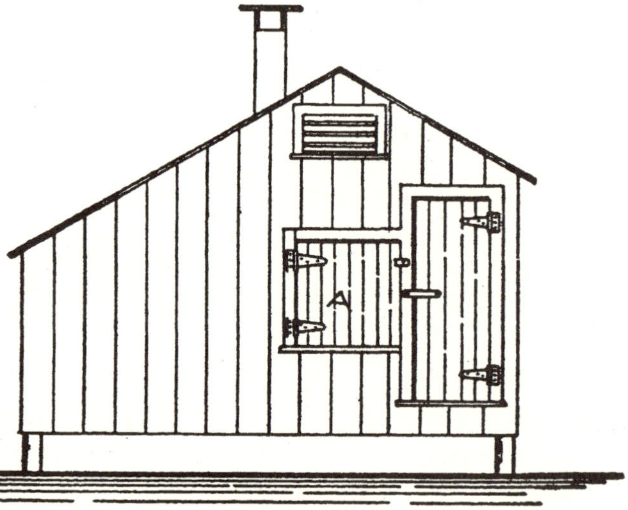

The south side of the building is only four feet high above the floor and the windows are placed well down. This has the advantage in the winter time of letting the sun shine on the floor where the chickens can make the best use of it.

The north wall is six feet high above the floor and this wall is made tight to keep out the cold. A partition three and one-half feet from the north side of the house forms an alley and the nests are placed against this partition so the eggs may be gathered without going into the henhouse proper. The nest boxes are placed high enough above the floor so the fowls may use the space under them for scratching. The nest boxes are easily removed for cleaning and they are covered with a steep slanting roof to prevent the hens from roosting on them.

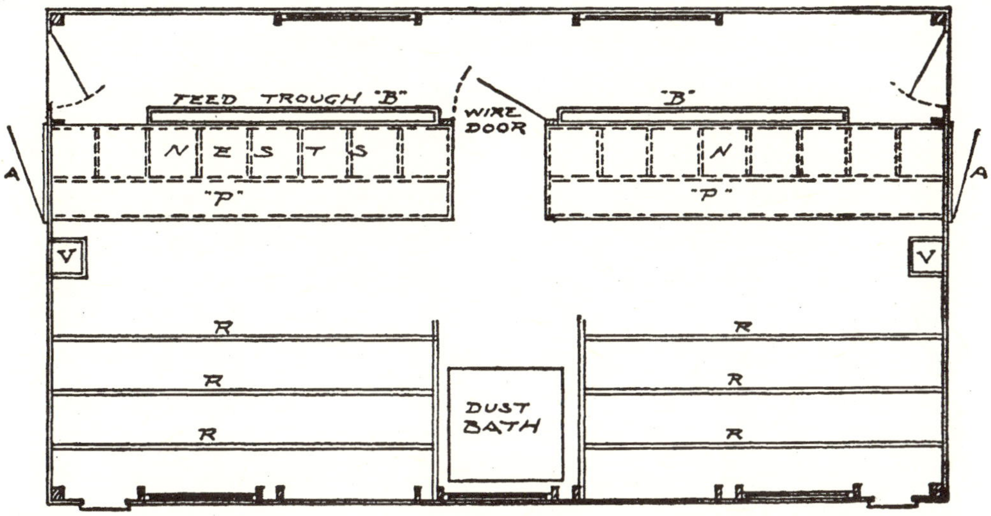

The ceiling in this house is an important feature. It is made by nailing one by six boards on the lower side of the ceiling joists about two inches apart. In winter the space above this slatted floor is filled in with straw for the purpose of having good ventilation without creating a draft. In the summer time the straw is removed and the place thoroughly cleaned.

For further ventilation there is a vent stack in each end of the building which comes down to within a few inches of the floor. These ventilators pass out through the roof and extend above the highest point and are capped to keep out the rain. There is also a slide near the bottom to regulate the amount of air. If heavy fowls are kept in this house good ladders should be provided to help them up and down or they may get bumble foot.

Ice Houses and Cold Storage

ICE FOR COLD STORAGE—A223

Blue Prints Will Cost, $15.00

ICE is still the most economical cooling medium for small refrigerators. Small automatic ice machines are in use but more or less expert supervision is necessary to keep them in working order. The advantages of direct ice cooling are simplicity and comparative economy, but the average refrigerator cooled by ice will hardly give a temperature below 45 or 40 degrees F and it has the disadvantage that the air is too damp for best results. Sensitive products like butter, milk and eggs will not keep well in a moist atmosphere.

To meet the difficulties between a large extensive refrigerator plant and the ordinary ice cooled refrigerator Madison Cooper has invented and designed what he calls the "Gravity Brine System" which will produce a temperature as low as 14 degrees F. The system is very simple and may be applied to a very small plant or one of considerable dimensions.

The diagram shows the principle upon which it works. In the small plant there is an insulated box or tank A on the floor above the cooling room. In this box is a coil of pipe which reaches down below into and coiled in the cold room B. This pipe coil is filled with a brine solution of chloride of calcium. The tank A is then filled with a mixture of crushed or broken ice and coarse salt which cools the brine. As cold brine is heavier than warm brine it settles in the pipes and the warmer brine from below rises in the other pipes to take its place. This keeps up a constant circu-

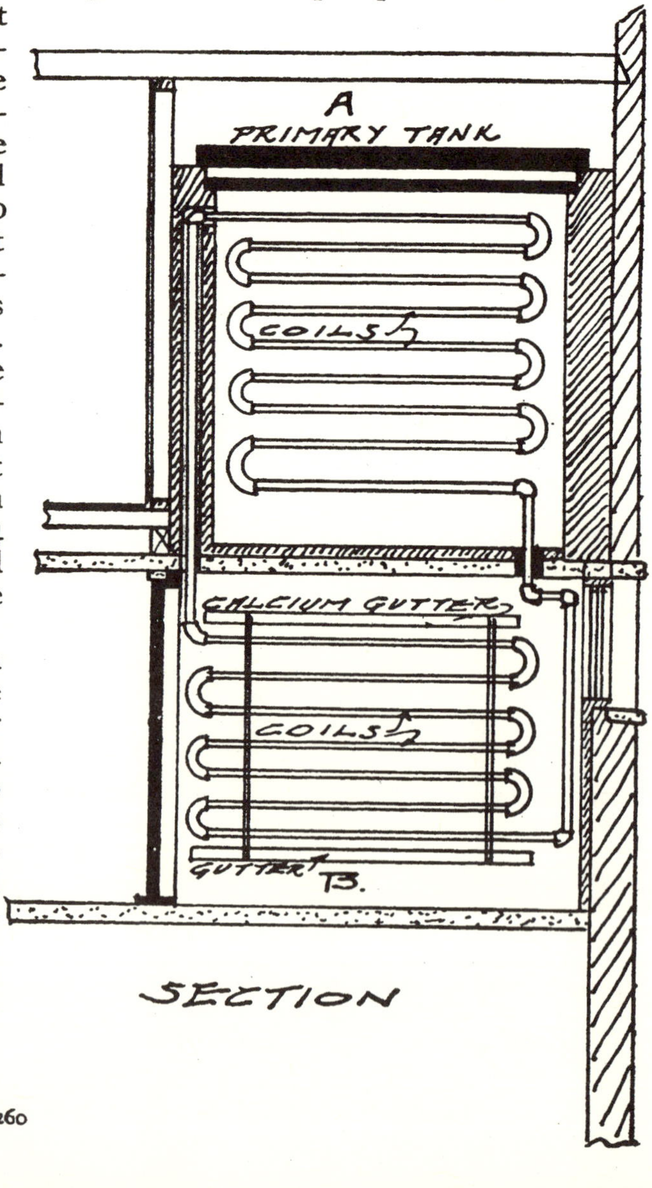

lation whereby the pipe in the lower room is kept cold and the degree of cold is regulated by the size of the cold storage room, the area of pipe surface and the amount of salt used with the broken or crushed ice. The arrows show the direction in which circulation flows. More or less pipe is used in proportion to the amount of cubic space to be cooled and the temperature maintained.

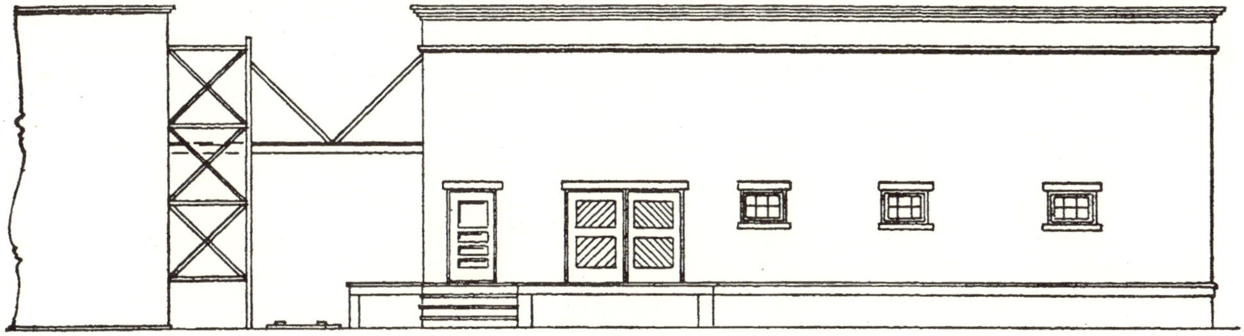

FRONT ELEVATION

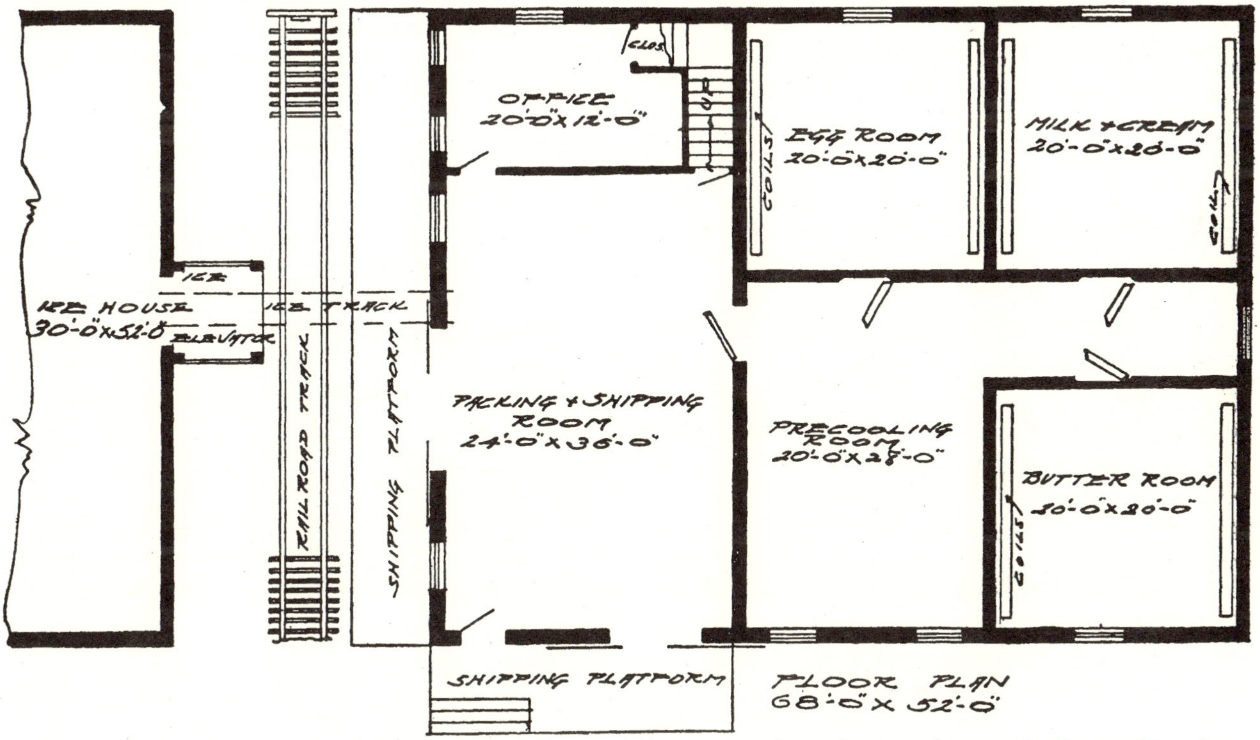

FLOOR PLAN 68'-0" X 52'-0"

Ice and salt will produce a temperature below zero, so the circulating brine may be easily cooled to five or ten degrees F, which, when the pipes are properly proportioned will produce a temperature in a good, well built storage room of from 10 to 40 degrees above zero. Mr. Cooper states that a temperature as low as six degrees has been obtained.

These temperatures of course are sufficiently low for all cold storage and freezing purposes in common use. In fact, thirty to thirty-three degrees is about as low as is required for ordinary cold storage.

It is well known that ice forms on cold storage pipes when the temperature is below the freezing point and the collection of ice not only interferes with the proper cooling process, but it adds dampness and this is to be avoided as far as possible. Mr. Cooper uses a process which he has patented called the chloride of calcium process for preventing frost on refrigerating surfaces and for purifying and drying the air of cold storage rooms. This process is extremely simple and consists simply of supporting a quantity of chloride of calcium in the lump form in perforated troughs or gutters above the cooling pipes of the Gravity Brine System in the cold storage room to absorb the moisture and carry it away.

The whole cooling system apparently is very simple and it looks as though any careful person of ordinary intelligence without special mechanical skill could look after it and keep it in good condition. This cooling system is in use in different parts of the country for the cold storage of butter, cheese, fruits, eggs, dressed poultry, meats and for the manufacture and keeping of ice cream.

Some of the plants have different rooms maintained at different temperatures. Some have fans for ventilating purposes and some are fitted with ice crushing machines, power elevators, electric lights, etc. It seems that the system may be employed to advantage for very small plants or for very large ones and that the results are equally satisfactory.

Farming communities within easy reach of railway stations should be supplied with a cold storage plant, and this plant should be under the management of a committee of farmers. Then it would not be absolutely necessary to ship perishable products when the market was unfavorable.

Cold storage, however, is a business proposition and to be successful it must be managed on business principles.

The cuts besides illustrating the principle of cold storage show how a practical cold storage warehouse may be built. The ice house is indicated at the left with an ice elevator to hoist the ice into the building from the cars. There is a plaform which connects the two buildings so the ice may be conveniently transferred to the upper part of the cold storage warehouse where it is broken up by machinery, mixed with salt and packed in the coils.

CHEAP ICE HOUSE—A222

A very cheap house will keep ice. All you really need is a roof to keep the rain off and boarding at the sides to hold the sawdust in place. If you put up a cube of ice in the winter time ten feet in diameter and keep it covered with a foot of sawdust all around you will have ice for farm use all next summer. This is not saying that a good, well-built ice house is unnecessary. The idea is that no farmer need do without ice because he cannot afford an expensive ice house.

The principle of keeping ice depends in the first place on getting quantity enough together to maintain a low temperature, then the drainage must be sufficient to carry away the water as the ice melts and drips. In the third place you must keep the air away from it which may be done by keeping wet sawdust continually close around the ice.

You can set up four poles, nail rough boards on the inside, leaving a foot space all around the ice and fill this space with sawdust. Then if you have a roof over it that will turn the rain you can have ice in hot weather, but such a building would be an eye-sore on the farm and a disgrace to the owner. Because you build cheap you need not build something you will be ashamed of afterwards.

This little ice house may be built very easily and cheaply and it will look right when finished and prove very satisfactory. There is a light framework of two by fours boarded on the outside by drop siding. The house is twelve feet square and is twelve feet high to the plates. The rafters are tied together by collar beams placed well up to leave plenty of head room. You will need head room in putting in the top courses at filling time. You can board up the inside or not; it makes but little difference. It is better to put the house in the shade of a large tree, or on the north side of a building, but drainage is more important than shade. If the drainage is not good naturally, put in tiling. Fill above the tiling with cinders and put a foot of sawdust on top of the cinders. Don't use straw if you can help it because it rots. Pine sawdust is the best if you can get it.

Fill the house in January or February. Throw water over the ice cold nights so it will all freeze solid together, then in March put on the sawdust, filling in between the ice and the sides of the house and cover the ice a foot or two in depth on top. Then as the warm weather comes on in spring climb in on top of the sawdust about once a week just to see that there are no cracks made by settling that will let in the air to melt the ice. Your summer supply may depend on packing the sawdust a few times during March and April.

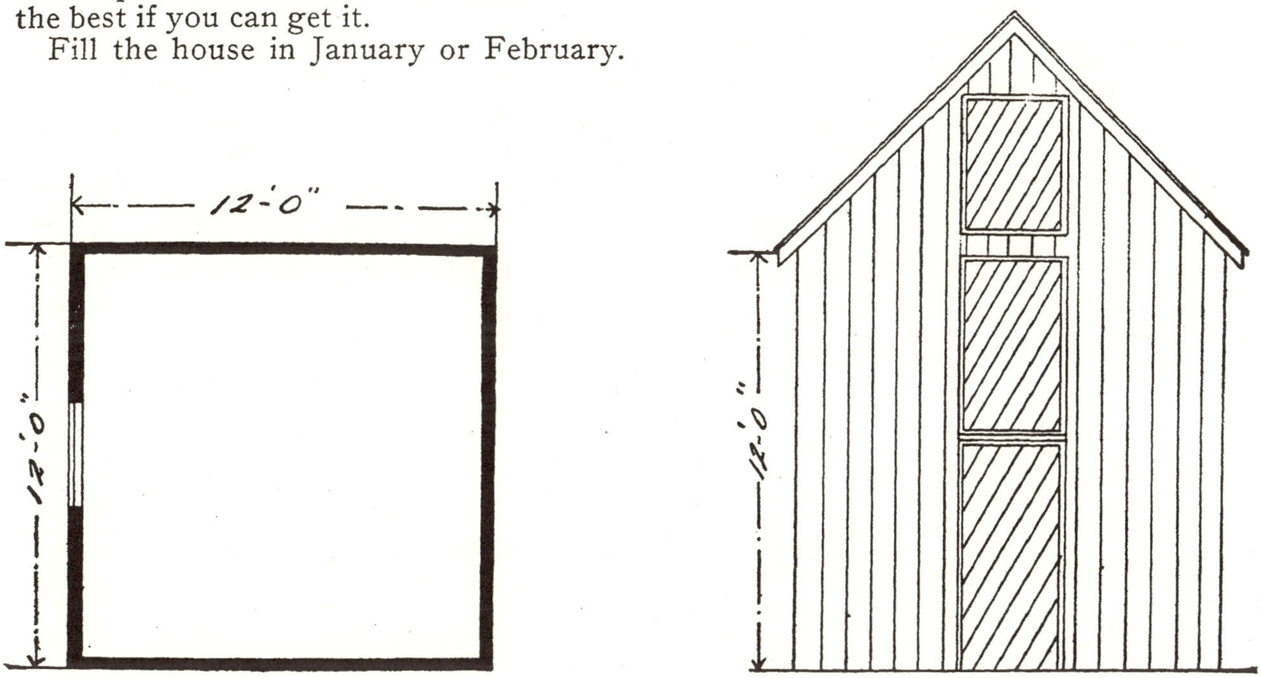

PLAN AND ELEVATION OF ICE HOUSE

LARGER ICE HOUSE—A224

Where a large number of cows are kept and a large quantity of ice is needed in the dairy every day in summer it will pay to put considerable expense on the ice house. The cooler and better the house the more economically you can keep ice. One builder even recommended one air space after another up to seven separate air spaces in order to prevent heat from striking through to warm the interior of the house.

The ice house should be nearly square because the ice keeps better when it is in one compact mass. In starting to build as good an ice house as this one it pays to put quite a bit of work on the foundation. Ice melts in the best houses and you must take care of the drip. This plan provides for a course of tiling all around the outer wall and a course of cross tiling every four feet, the cross tiles connecting with the outside tile and all empty in the one outlet through a trap at the lower corner.

No earth is put back into the excavation, but the tiles are covered with coarse cinders, cinders that have been raked over and only the hard burned parts used. If the cinders are too fine for a rake, use a screen, but manage some way to keep the ashes out. Sometimes the ashes pack down and hold water like cement.

The house is sixteen by twenty feet and the wall is four inches larger each way. The sills are made of two by eight double and are laid flatways on the wall in soft cement. The wall is also smooth up to the sill on the inside with cement neatly troweled around.

Two by four studding is set flush with the inside of the sill. Building paper is put

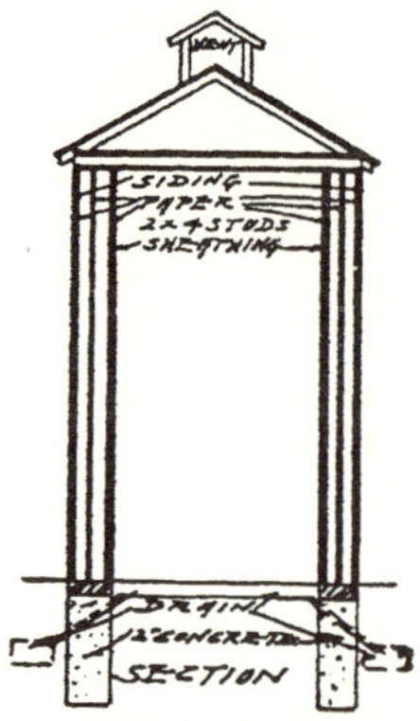

on the outside of these and another set of two by four studding set against them outside of the paper. In this way a wall eight inches thick is made with paper partition in the center.

The wall is then finished with paper and siding outside and with paper and siding inside, put on as carefully as possible and made as nearly air tight as good material and good workmanship can make it. The paper and boarding is lapped over on the sill at the bottom and on the plate at the top so that the sides of the building are completely insulated from the sill to the ceiling.

The ceiling is made with two by six joists and matched boarding laid with paper like a floor on top and another matched boarding nailed like a ceiling with paper on the under side, the paper being tacked to the joists both above and below. Paper also is used on the rough boards under the shingles and a good ventilator is provided to keep the air moving in the attic.

Because of the ceiling it is more work

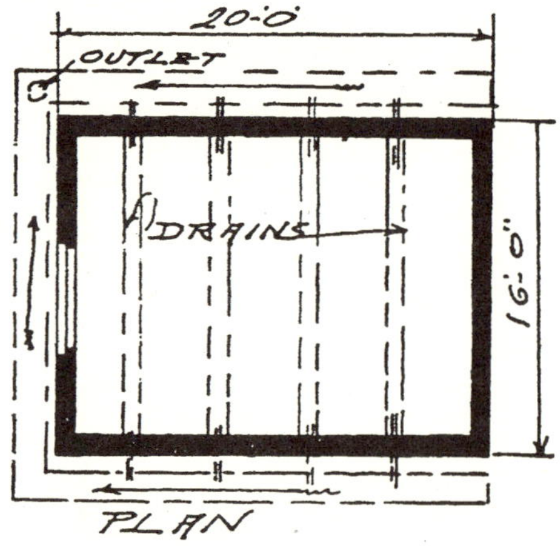

to put in the ice when you get near to the top, but an ice room built in this way is a refrigerator in itself and it will save ice enough to pay for a little inconvenience at time of filling. The doors fasten in front like refrigerator doors. They are built like the sides of the house with air spaces and are made to fit tight in the jambs. All doors and door frames must be made of kiln dry material and filled to resist moisture. There is no floor, the sawdust just rests on the cinders, and the ice rests on the sawdust which should be at least a foot deep on top of the cinders.

TWO HUNDRED TON ICE HOUSE—A228

An ice house to hold two hundred tons of ice is given in this plan. This ice house was built on a large dairy farm near a good sized village. Some seasons the farmer sells considerable ice to the village at paying prices.

The building is twenty feet wide by thirty feet long and sixteen feet high to the eaves. When completely filled it would hold about two hundred and twenty-five tons.

The exterior is finished with drop siding and a stained shingle roof. Next to the siding is a layer of building paper, inside of this and nailed to the outside row of two by four studding is matched ceiling of good quality. Then comes a dead air space four inches thick. Next is a layer of hair felt seven-eighths of an inch thick nailed to the inner edges of the four-inch studding; inside the hair felt is another matched ceiling of narrow pine sheathing,

DESIGN OF AN ICE HOUSE HAVING A CAPACITY OF 200 TONS

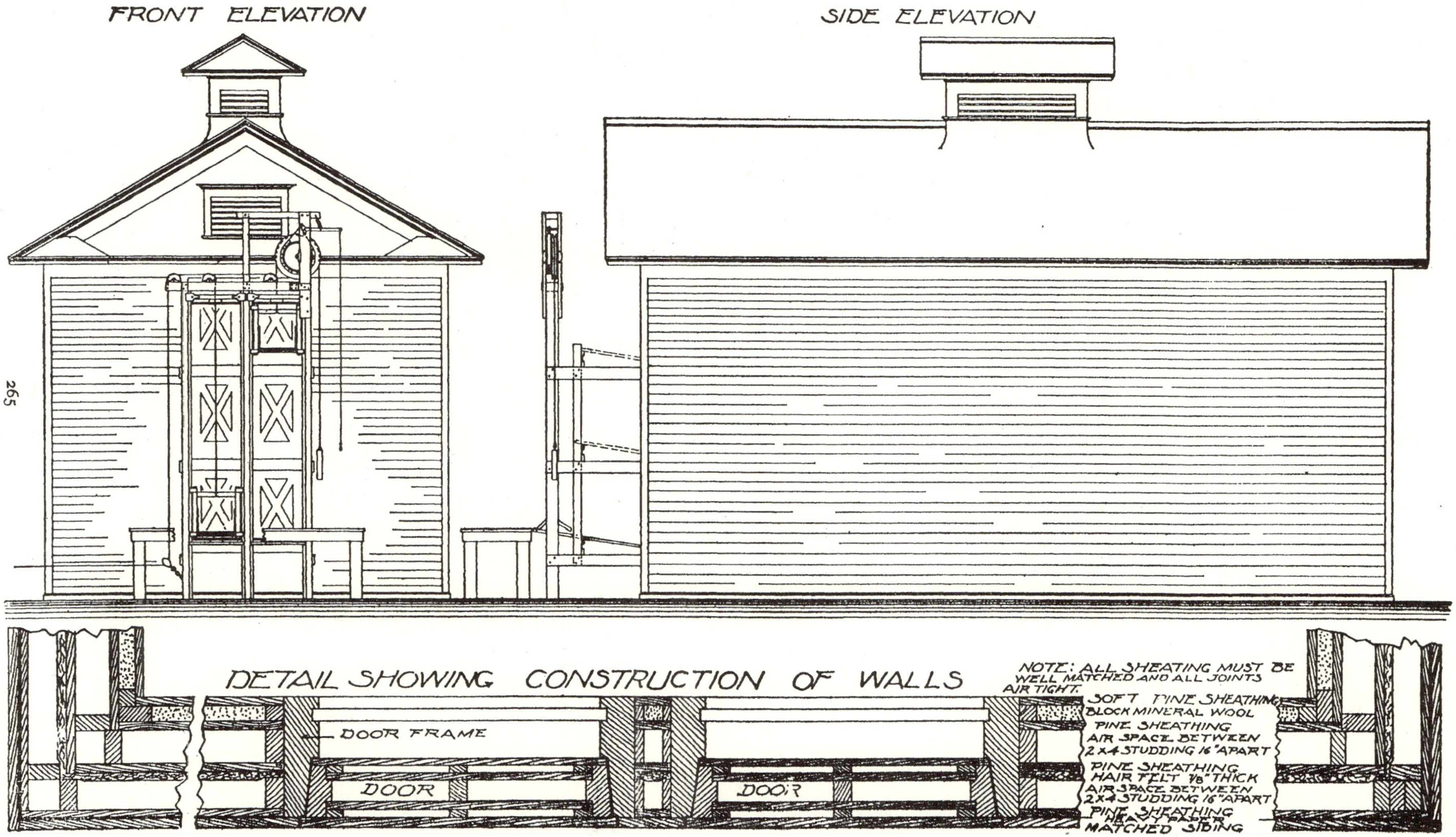

then another row of two by four studding lined on the inside again with another boarding of matched pine sheathing, then an inch of block mineral wool, and this is protected on the inside with another boarding of matched soft pine sheathing nailed to furring strips. All this work is very carefully done to prevent so far as possible the slightest air connection between the different spaces. It is recognized that a dead air space is the best possible non-conductor of heat or cold.

There are six doors and they are just as carefully made as the siding. The detail drawing shows how they are fitted. Inside of the doors the opening is further closed and sealed by a double thickness of loose inch boards, which fit into the grooves and are laid to break joints. These boards are put in place as the filling proceeds and are taken out one at a time as the ice lowers in summer.

The ceiling over the ice is just as carefully constructed as other parts of the building and the space over the ceiling is kept cool by a ventilator in each gable end and another ventilator in the roof.

All these details are very important but they are not more important than the covering for the ice, which should be of saw dust if possible to get it.

An interesting feature of this house is the simple elevator to be used in filling. It is a double gig elevator so arranged that one gig goes up as the horse walks in one direction, and as the horse walks in the other direction the first gig lowers and the second one goes up. Perhaps this is the quickest arrangement made for the purpose, considering its simplicity.

REFRIGERATOR ICE HOUSE—A118

An ice house with a cold storage room is shown in plan (A118). The walls are built hollow with paper inside and out.

In the cold storage department there are several thicknesses of paper in the inner wall to make the dead air space as tight as possible. If you have ever undertaken to make an absolutely dead air space you understand the difficulty, or the impossibility of doing it. There is sure to be a crack somewhere to let the air through, but this plan probably comes as near to it as is necessary.

When an ice house is made as tight as this it is necessary to let the top air out. For this reason a ventilator is built in the roof to encourage a circulation of air between the upper ceiling and the shingles.

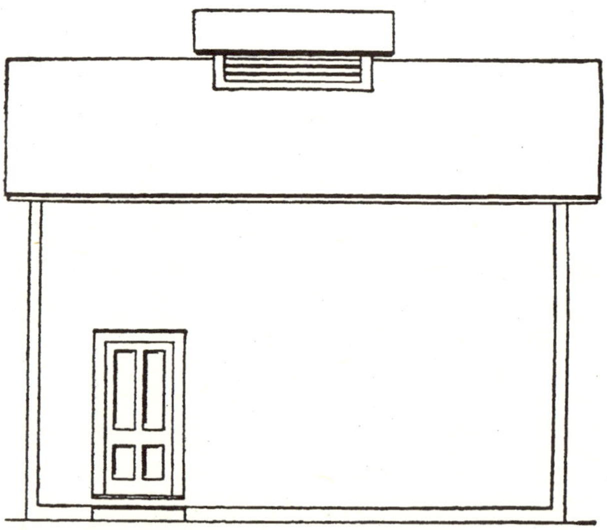

FRONT ELEVATION

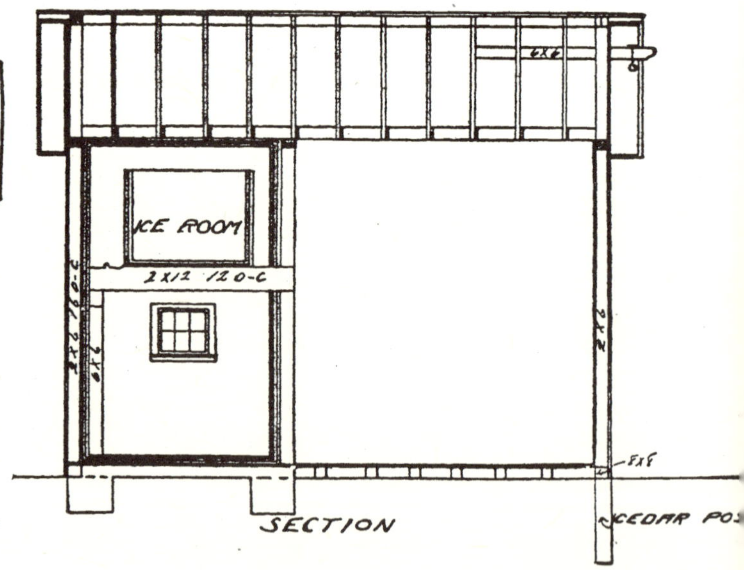

SECTION

In this arrangement the cold storage department is supplied with ice as needed by putting in a quantity, say once a week.

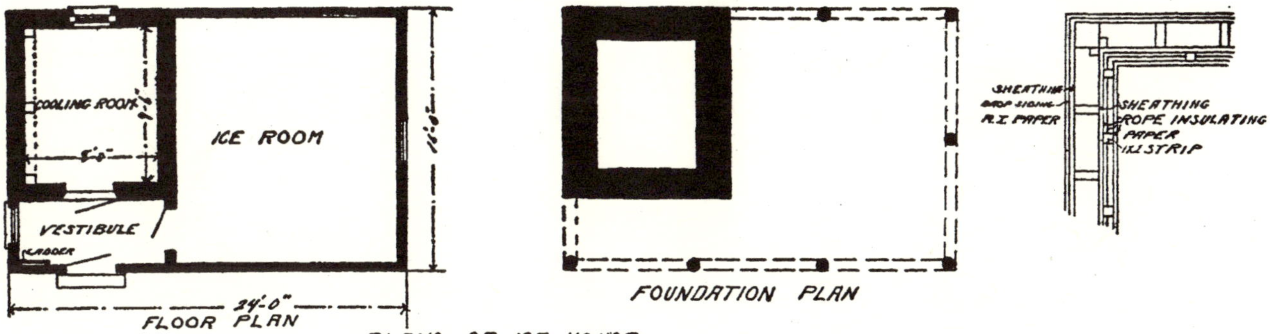

PLANS OF ICE HOUSE

The construction of an ice house like this requires good workmanship. You will need the best mechanic in the neighborhood and it will pay to read up on cold storage before you start in. If it is made just right it will be a great comfort and satisfaction, but if it is not made right it will cause a great deal of trouble and be a continuous annoyance.

Miscellaneous Farm Buildings

IMPLEMENT SHED—A148

AN implement shed sixteen feet wide by forty-eight feet long is given in plan (A148). This shed really is built in sixteen foot sections and may be carried to any length, but this size will hold the implements and machinery on an ordinary farm and leave room at one end for a work bench and repair shop.

The front is all doors so that any part of the shed may be opened and any implement removed without the work of getting it past some of the others. We have all had experience in crowded quarters for farm machinery. We have been obliged to call all the men together and occasionally invite the women to help get a grain drill out from behind harrows, plows, cultivators and other machinery. One reason why farm machinery is neglected is because farmers have no place to keep it. It is not repaired when it should be for the same reason. It is quite a job to do a simple piece of repair work if you haven't the tools or the room in which to do it, but anybody can clean up machinery and oil or paint it if they have a comfortable place to work and the tools to work with.

ELEVATION

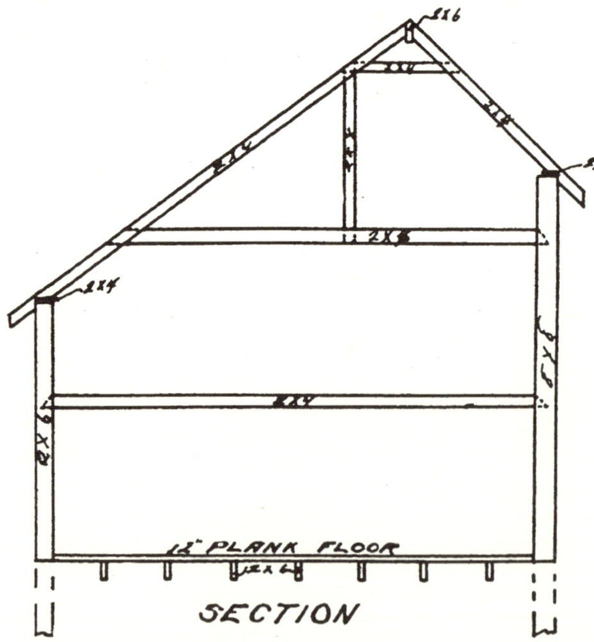

SECTION

The front part of this shed is built higher than the back part in order to leave head room. If you want to get in with a binder with the reel on, or to house a threshing machine or traction engine you need about ten feet to the top of the doors, in the detail drawing are intended for the end bents only.

In the end of the shed most convenient a good solid bench should be rigged up

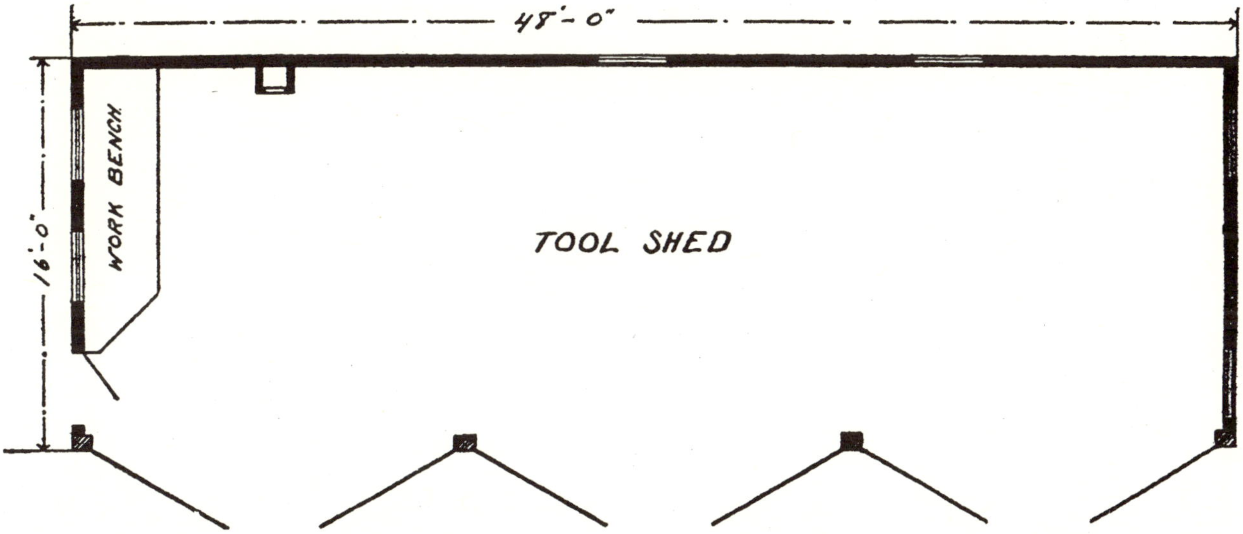

but you don't need so much height to the back end. The doors in this plan are ten feet high and the cross girts are the same height because it is sometimes necessary to move the machines lengthwise of the shed and the same head room is then needed. A truss is formed at each bent with the rafters to prevent the building from spreading. The two by four nailers shown and fitted with a good vise. At the back of this bench there should be a long low window similar to those used in blacksmith shops all over the country. The bench should be heavy, solid and at least three feet wide. There should be a good floor especially in the bench end of the building and it is a good plan to put up a chimney and have a stove there.

OPEN VEHICLE SHED—A221

A shelter for vehicles in this case is provided by a row of posts covered by a protecting roof, braced as shown in the drawings. It is necessary to brace a building of this kind and as bracing near the ground would be very much in the way this form of truss work has been adopted.

Eight by eight wooden posts are set in a double row sixteen feet apart both ways, thus forming a series of bays sixteen feet square. The wooden posts are set in concrete abutments twenty inches square and four feet high set in the ground from three and one-half to four feet according to the nature of the ground.

These abutments are made by filling the holes, which are dug the exact size, with concrete tamped in in the usual way, but the tops are made uniform and beveled as shown by the use of a box form hinged at one corner and fastened with a hasp and staple at the corner diagonally opposite.

The holes are filled in about a foot deep with concrete, then the wooden posts are set up carefully and plumbed and staylathed to keep them in position. Then the concrete is filled in around them and the abutments carried a foot or so above ground to form a hub guard.

The drawing shows a shed with five

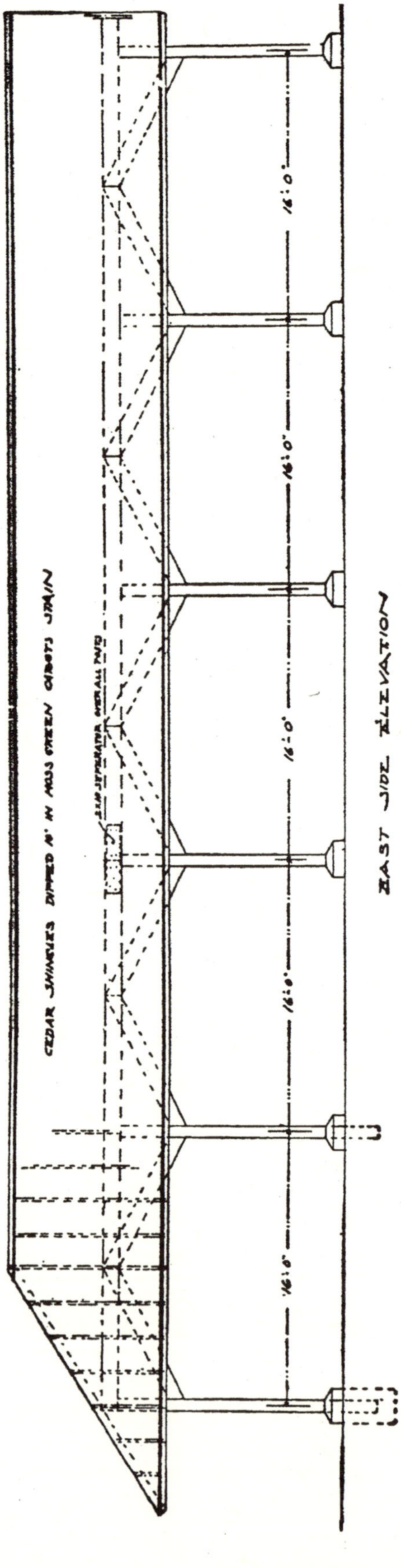

of these sixteen foot compartments, but fewer can be used if a smaller shed is required, and the building would look alright when finished.

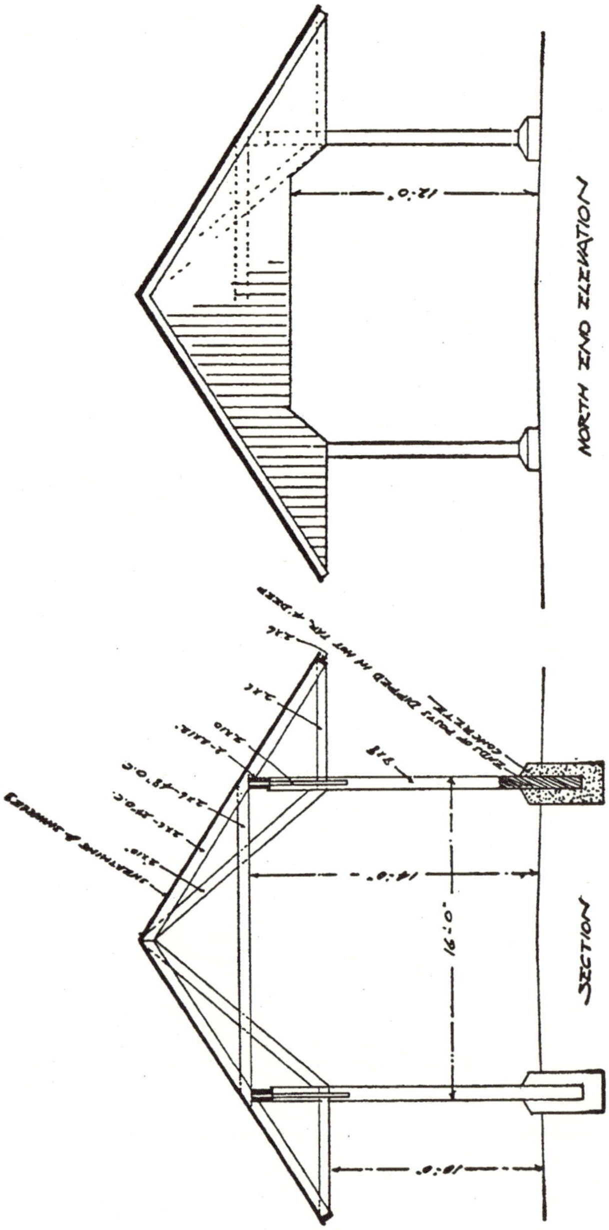

The entrance at the ends is twelve feet high, which will permit driving in with a load of hay or sheaves. A shed built like this would work well as a stock shelter.

SMALL WAGON SHED—A108

A wagon shed twenty feet wide and forty feet long like the one in the plan illustrated, is a useful building on every farm. One thing is important about a wagon shed, and that is to have the entrance wide enough to get things in and out easily and quickly. This double door

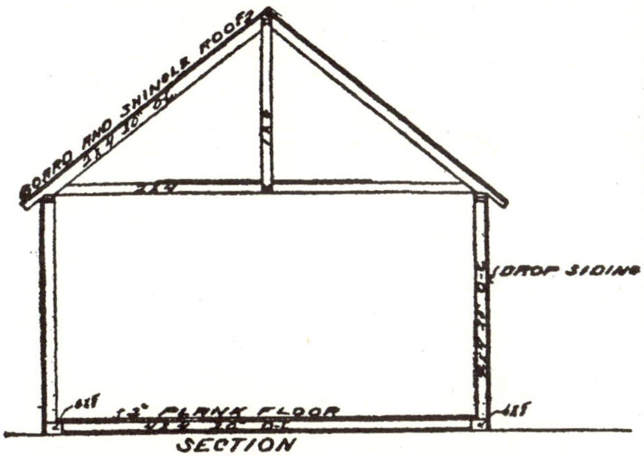

SECTION

FLOOR PLAN OF WAGON AND IMPLEMENT SHED

gives an opening ten feet wide, which is very good for small implements, but some binders require about sixteen. The door entering an implement shed must be high enough to let in the highest implements used on the farm, and there must be no cross timbers inside lower than the top of the door. A binder with a reel on takes considerable room.

The farmer building the shed will know whether he wants to house a binder under full sail or, whether he wants to take it apart, and will, of course, build a doorway accordingly.

Implement sheds like all other buildings should be designed for what is to be required of them. An implement shed is a necessity on every farm, but some farmers want to house threshing machines and traction engines, while others want a shed to hold mowers, plows, cultivators, a wagon or two, and perhaps a few barrels and other truck. A large building, of course, would answer for everything, but it is not necessary to build bigger than a man wants.

A good many tool houses are built without floors, but the extra cost of the floor is more than offset by the dryness and freedom from rust on the machinery. Wagons and machinery require repairing, which is easily done in a building like this when you have a good floor to work on. Odd days in winter may be profitably spent in such a building with a few carpenter's tools, a paint brush and an assortment of paints and oils.

CHEAP HOG HOUSE—A122

The cheapest kind of a hog house is shown in plan (A122.) It is only seven feet six inches wide, but it may be any length. This house is thirty-one feet six inches long because this length is covered by two sixteen foot joists. Even on well regulated hog farms where there is a good solid hog house this shed affair will be found useful to hold the overflow. It often happens that shoats in fall are kept in a muddy feed lot or sold too soon for lack of just such shelter as this to hold them while being finished. Beginners in the hog business could not do better than to build a little cheap hog house like this to start with.

The seven and one-half foot width permits of roofing the shed with sixteen foot boards cut in two in the middle. Each sixteen foot section will make two pens nearly eight feet square which will hold from five to seven or eight pigs according to size. The posts are just set into the ground and the floor raised about a foot to keep it dry. Four two by six joists run lengthwise of the building and the floor boards run crosswise and slant back for easy cleaning. A space is left between the floor boards and the boarding at the back so a scraper may be used. In cold weather this space is closed by a hinged board which drops down on the inside. This precaution is necessary because a cold draft on the floor is a very bad thing for hogs. This little hog house don't run into very much money but it is a very useful, practical affair.

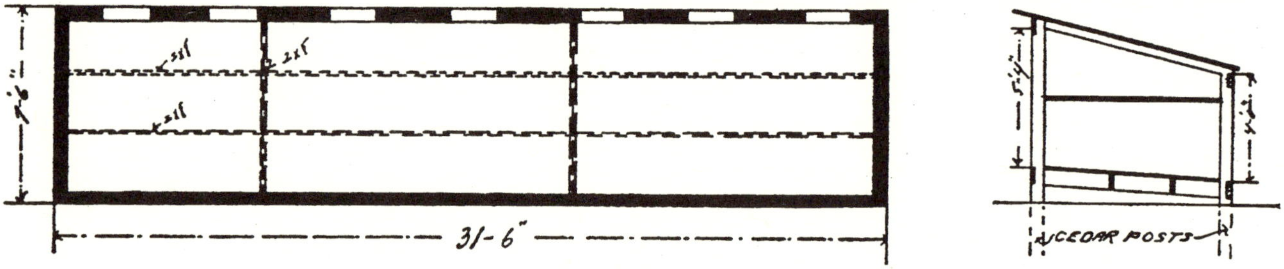

PLAN AND SECTION OF HOG HOUSE

HOG HOUSE—A109

In building a hog house it is necessary to consider convenience in getting the hogs in and out, to provide means for loading them into wagons and a place for heating water and to do the work of killing. This plan offers an opportunity to back a wagon up to the rear door for loading and a room in the front end away from the pens is arranged for a feed room and slaughter house.

Provision is made for removing hogs from one pen to another by having cleats in the alley for holding sliding doors.

Hogs thrive better when animals of the same size are penned together. Some grow faster than others and it is sometimes desirable to select out one or two from certain pens. That is the time when the alley door will be appreciated. Another

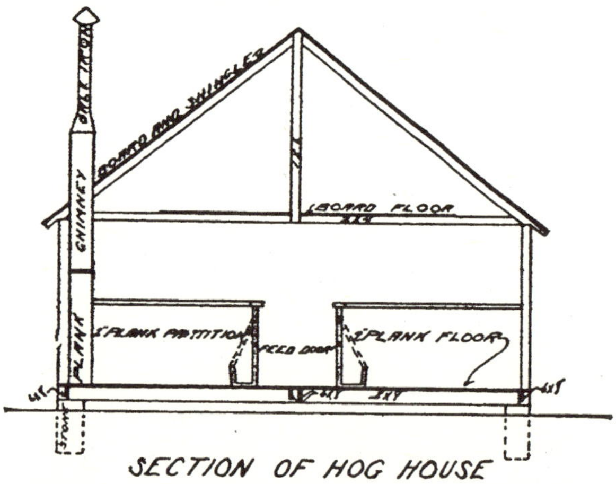

SECTION OF HOG HOUSE

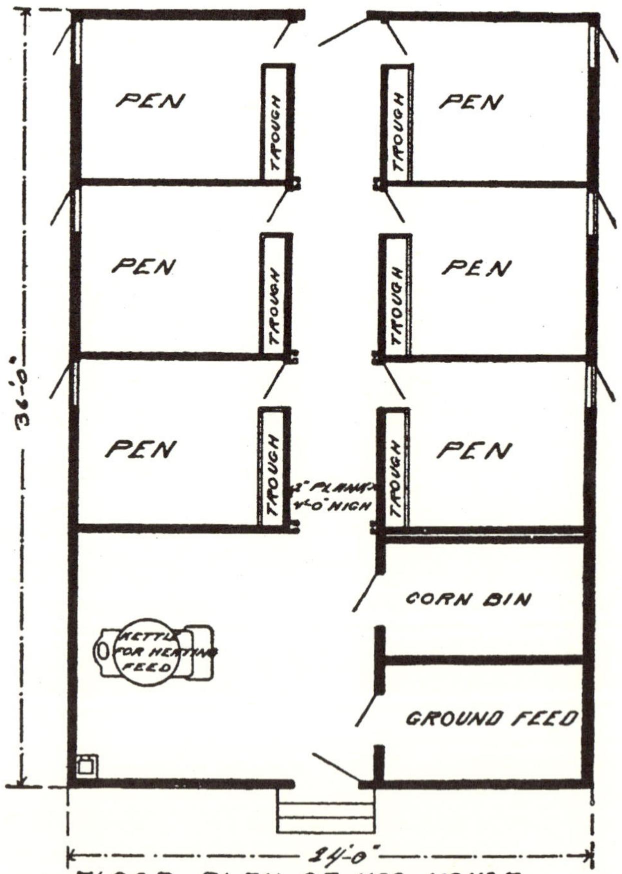

FLOOR PLAN OF HOG HOUSE

good thing about this hog house is the swinging front of the pens which swings back over the trough and prevents interference when putting in the feed. The partitions next to the feed room run to the ceiling but the partitions between the pens are only four feet high.

There is no cornice to the roof. The openings above the plates between the rafters are left for ventilation. This hog house will accommodate about forty hogs. From six to eight in a pen are enough, if more are housed together they pile up and smother each other.

HOG HOUSE AND CORN CRIB—A140

Hogs and corn may both be kept in the same house economically by building a house like the one shown in plan (A140). in the usual way with an alley between.

The floor above to hold the corn slants each way from the center. There is about

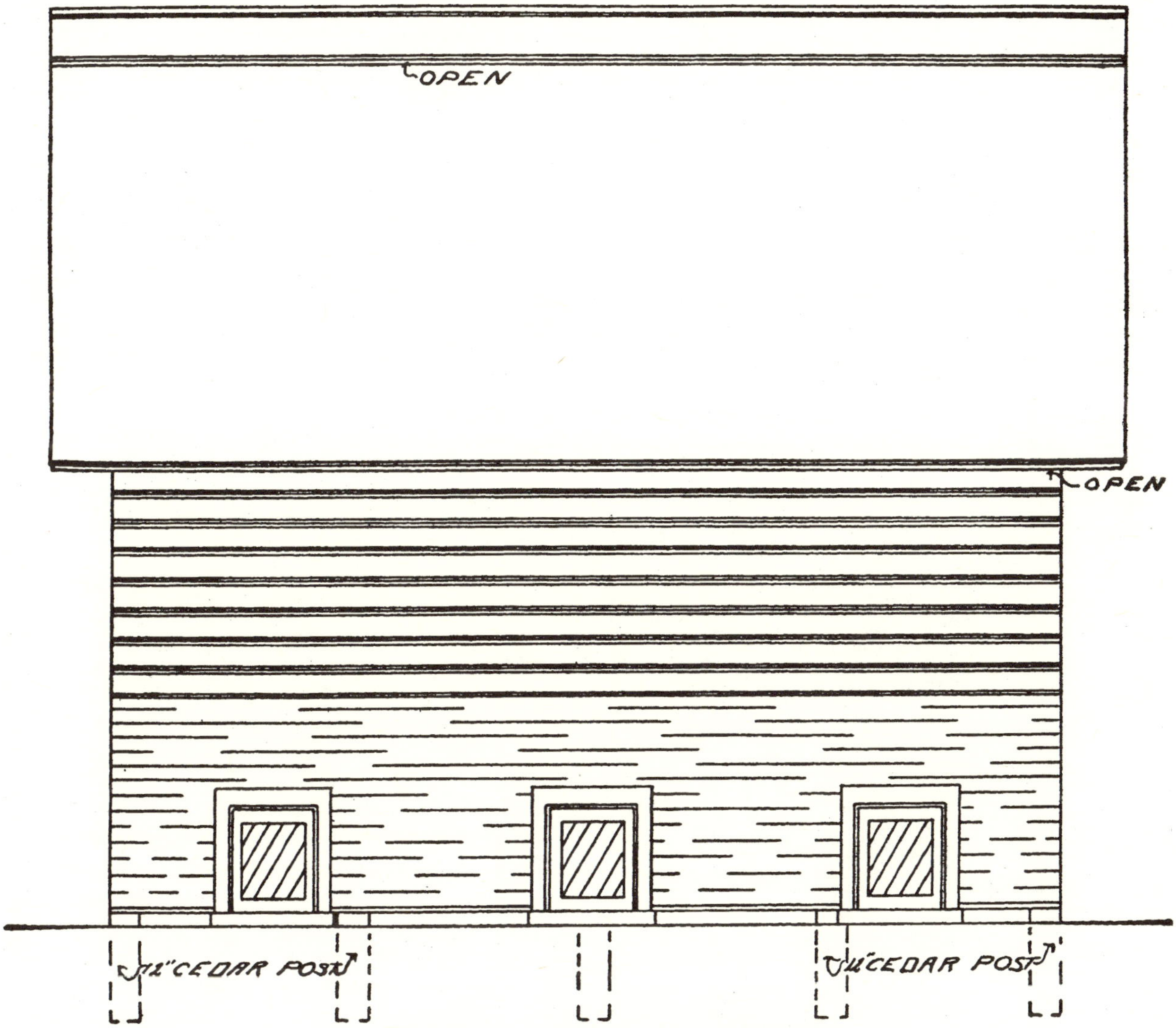

SIDE ELEVATION

The building is set up from the ground about a foot on posts and pens are made seven feet head room in the middle over the alley and the floor slopes to about five feet to the sides of the building. This is for two reasons, to get the corn down as low as possible and to divide it into two parts to prevent moulding. It is also necessary to put a slatted partition on both sides of the floor ridge if the house is filled full of corn. There are two windows in each end and the hog doors are hung with pins so they swing either way and the hogs open them going or coming. A pin at the bottom outside holds the door shut when it is desirable to keep the hogs confined.

Because of the shape of the corn floor it is necessary to support it well in the center which is done by running the alley partitions up to the floor joists above. This is very important because the weight of the corn will shove the sides of the building out if the floor is permitted to settle.

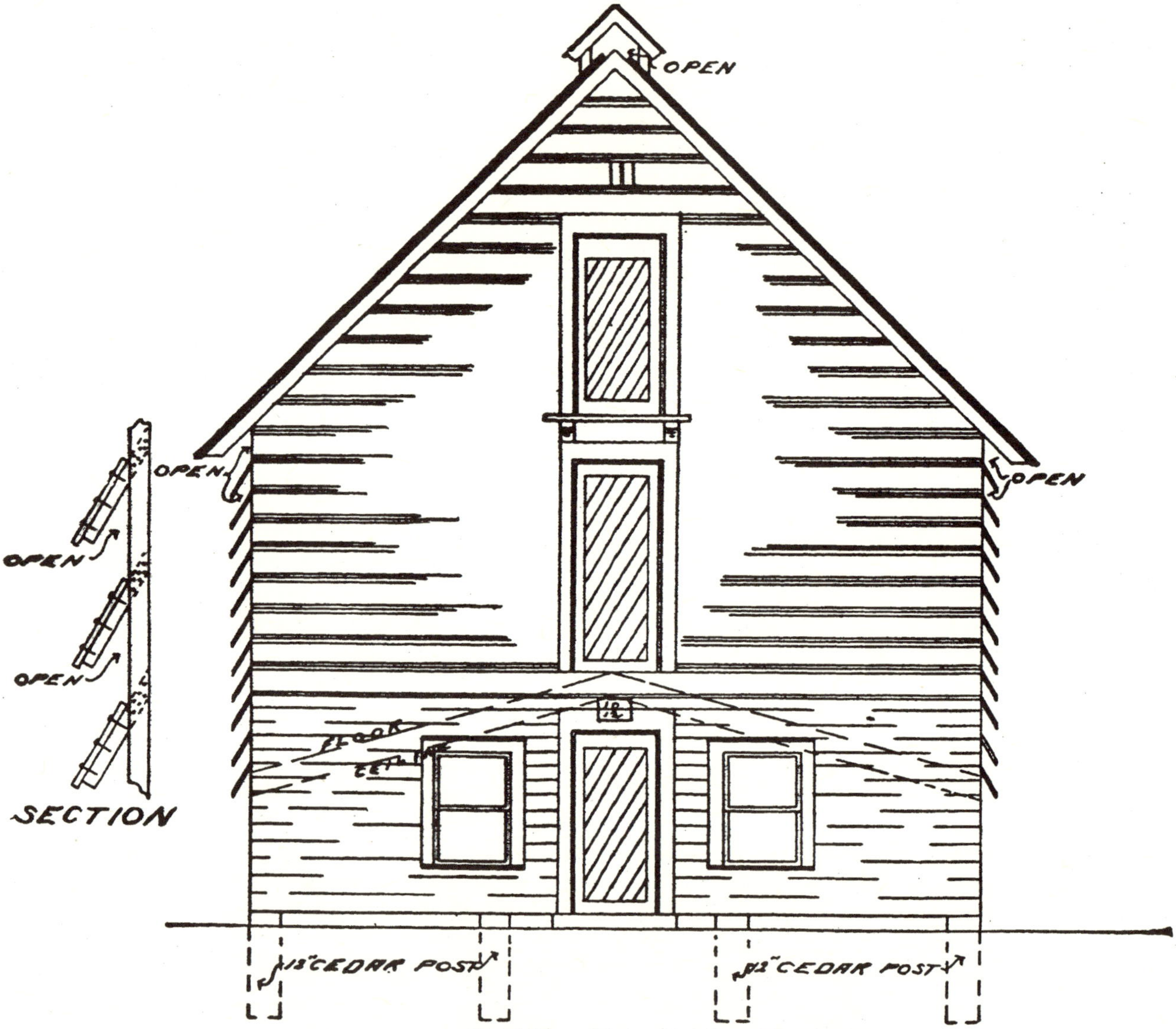

END ELEVATION

DOUBLE CORN CRIB—A105

An old fashioned style double corn crib with a drive between and a roof to cover both cribs is shown in plan (A105). This crib is set on cedar posts planted three and one-half feet in the ground and set up two and one-half feet above ground to be out of the way of mice and rats. The space

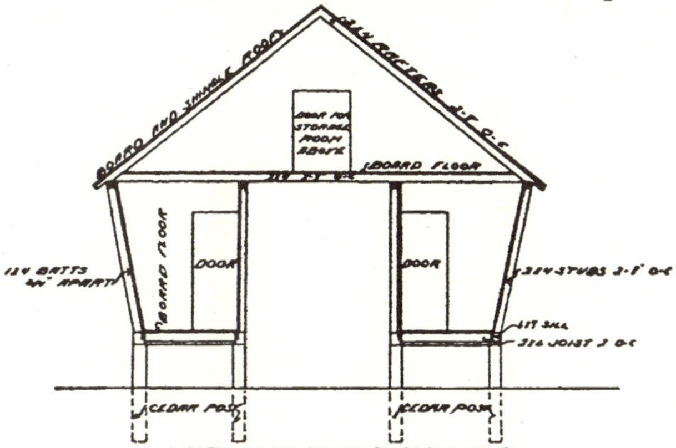

SECTION OF DOUBLE CORN CRIB

between the two cribs makes a convenient place to store a couple of wagons. The doors being at the end, the center space is left free for this purpose. A good many cribs built on this plan have the doors inside opening from the center passage, but if the space is desirable for wagon storage the doors are built at the end as shown. The storage room overhead will be found useful on any farm.

FLOOR PLAN OF DOUBLE CORN CRIB

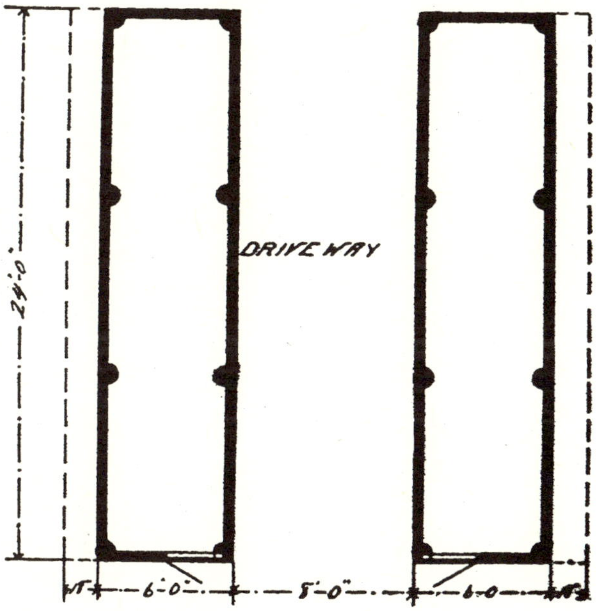

ANOTHER DOUBLE CORN CRIB—A120

A double corn crib with a storage room overhead and a driveway in the center is shown in the illustration. A peculiar feature of this plan is the siding which is ripped out it is run through a sticker to give the curve as shown in the detail drawing. This is an extra protection against beating storms and it is supposed to en-

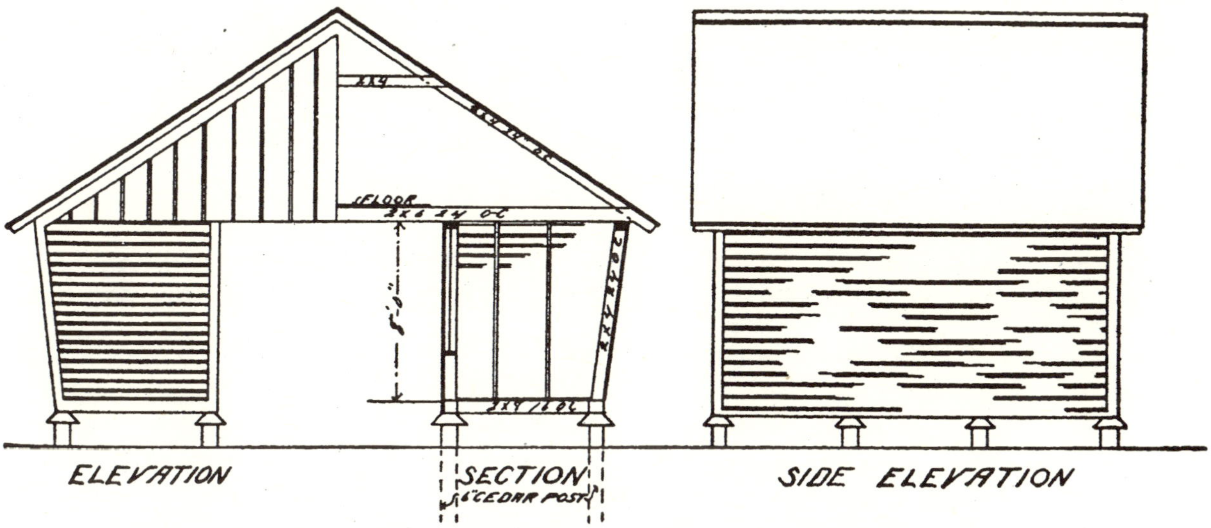

split from two by fours with a band saw in such a way as to get three pieces of siding from one strip. After the siding is courage a draft of fresh air up through each opening for the benefit of the corn. The strips are nearly an inch thick on the lower edge making them strong enough to hold the corn by being well nailed with wire nails on the outside of the studding which is placed twenty-four inches apart. It is impossible to get corn enough in a crib of this height to break the slats or shove them out. Corn cribs should not be more than six feet wide because corn will mould in a crib that is too wide.

DETAIL OF SIDING

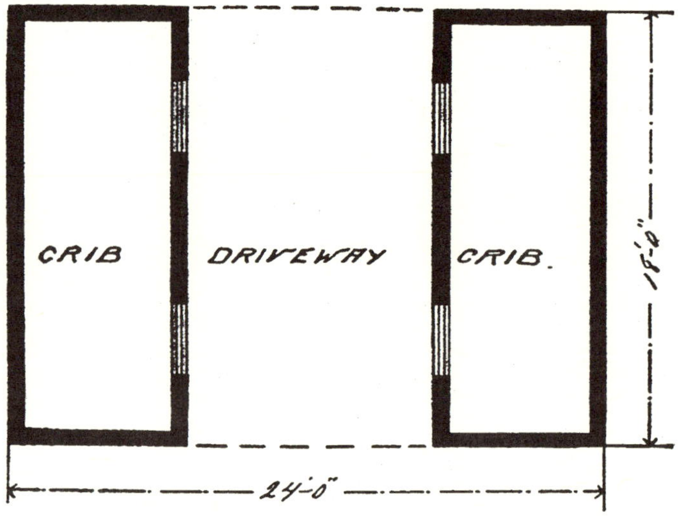

PLAN OF CORN CRIB.

The driveway in the center of a crib like this is very useful. There is room for a wagon or two and there may be pegs to hang a great many farm implements such as neck-yokes, extra whiffle-trees, chains and hand tools of all kinds. The loft overhead makes good storage for lumber and there is no better place for seed corn than to hang it by wires from the collar beams. The tin pans turned upside-down over the tops of the cedar posts will bother the rats most of the time, although they sometimes find a way to get in. Probably carelessness in leaning something against the crib helps them up in the majority of cases. Rats and mice are often carried in the crib with the corn. They are sharp enough to get into a bushel crate and stay there until they are carried inside. In this way a farmer often populates his own corn crib with rats or mice without intending to.

ROUND CORN CRIB—A142

So far as the size is concerned there is more room in a round corn crib than in any other shape made with the same amount of material. The building is easily constructed because it is all plain straight work and it is rat proof because it is set up two feet from the ground on cement posts.

The posts are made by digging holes in the ground three and one-half feet deep and about eight inches in diameter. Lengths of eight inch pipe made of galvanized iron are used to carry the cement two feet above the ground. Before commencing it is necessary to strike a common level at the surface of the ground so that when the pipes are all set up the tops of them will be the same height. The post above the ground and the post underground should all be made at the same time so that the cement will unite into one solid post.

The floor plan shows the way the joists are laid and the circles represent the girts to which the 1x4 upright pieces are nailed. As the crib is sixteen feet in diameter it is necessary to have a ventilator in the

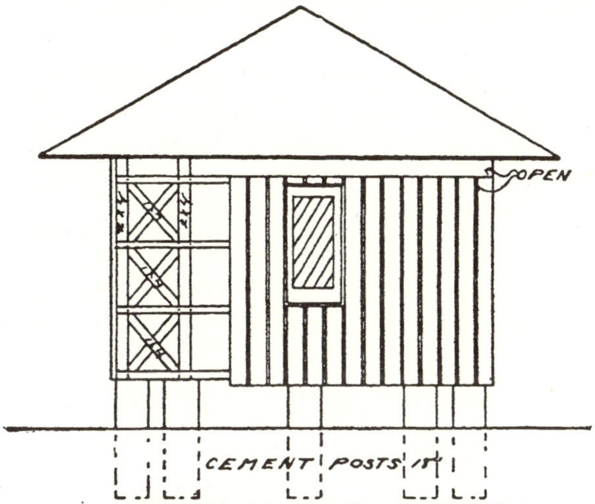

SECTION AND ELEVATION OF ROUND CORN CRIB

middle. Ordinarily it is not advisable to have a body of corn more than six or seven feet in diameter. By making the inner circle three feet we have six and one-half

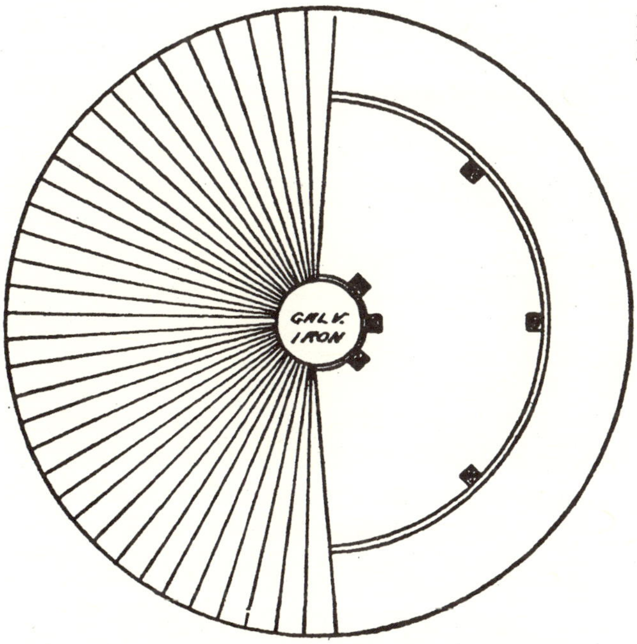

feet between the inner strips and the outer strips and as there is no floor over the joists in the center the air can pass up through the three foot ventilator easily.

The round girts may be made in two ways, either by using thin stuff and nailing one layer upon the other, breaking joints, or they may be ripped out of two

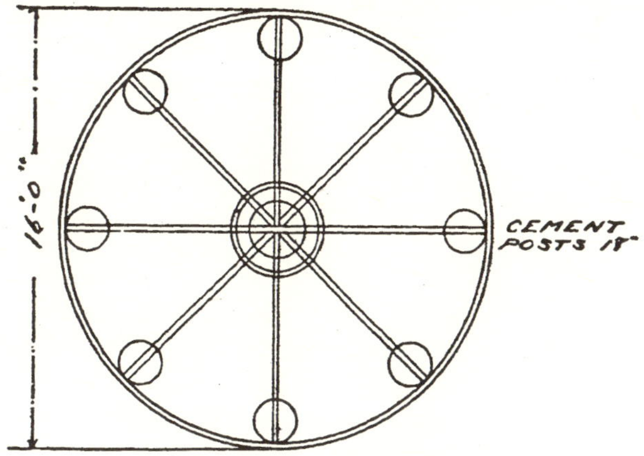

FLOOR PLAN OF ROUND CORN CRIB

inch planks. If ripped out of planks a single saw-cut through each piece of plank will shape the sections, cut as shown in the diagram. Use two inch plank ten inches wide cut to four foot lengths. Make segments enough to build up all the girts necessary by ripping the short planks lengthwise through the middle, then rip

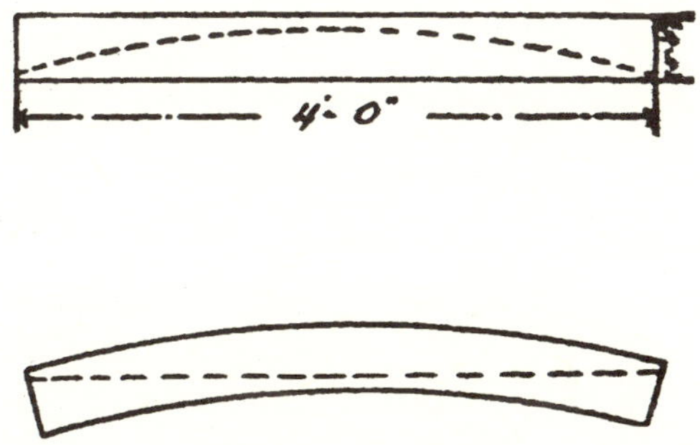

DETAILS OF ROUND CORN CRIB

again on the curved line. The finished girts are about 4x4¾ inches. There is very little waste. The roof is supported by a similar girt and this upper girt or plate is supported by extending some of the one by four pieces above the others as shown in the drawing. These extension strips may be doubled or two by fours used at these places. The crib is twelve feet high to the plate. An air space is left all around and this air space is big enough to shovel corn through. Of course the corn is put in at the door and at the opposite window until the crib is pretty well filled.

The roof itself is a very simple affair. It is supported by the plate and the ventilator shaft. The roof boards are 12 feet long and cut 11 inches at the wide end and 1 inch at the upper end or narrow end. These boards are nailed in place and the cracks battened. The center is easily filled in with sheet of galvanized iron having a cut reaching from one edge to the center. Such a roof if kept painted will last a long time. It is very light, cheap and easily made.

SINGLE CORN CRIB—A106

Sometimes a single corn crib is preferable to a double one. The corn keeps better in a single crib because the air circulates all around. Sometimes corn will mould in the center, even in a good crib that is properly constructed and not too wide. Sometimes farmers bore the floor full of holes to help the ventilation but this lets the shelled corn through and as dirt settles at the bottom the holes get easily covered over, and it is doubtful if they help very much. A better plan is to have the sides carefully constructed and to have the corn in a good condition when it is put in crib. A crib built after this plan may be any length but the posts should not be more than eight feet apart.

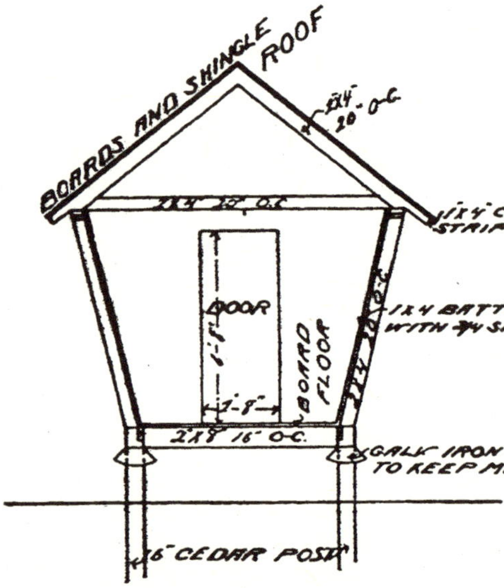

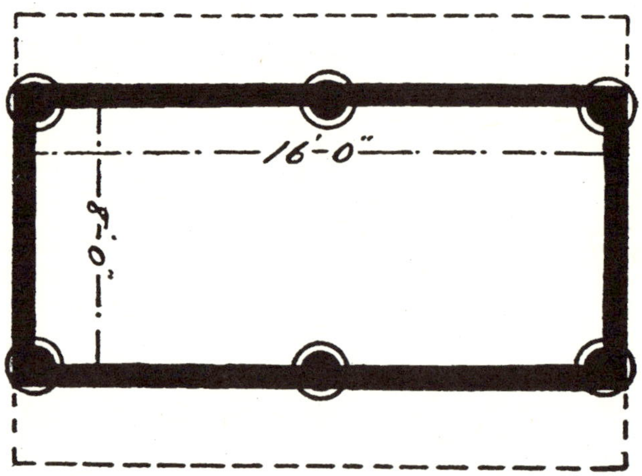

CHEAP SMOKE HOUSE—A149

It is not necessary to do without a smoke house on a farm. A small building that will answer the purpose may be had with very little outlay. The plan (A149) shows a little wooden smoke house eight by ten feet with sides eight feet high. It is big enough to hold as many hams and shoulders as farmers' families usually require with once filling, but it is an easy matter to fill the house the second time if you have the meat.

This little house requires no frame work at all. All you need is a four by four for sills and a two by four for plates and some more two by fours for rafters. You can even dispense with the rafters, except the two end pairs, if you want to make a board roof. It is better however to make a good shingle roof, then you have something that will last as long as you want it. For boarding you just take sixteen foot boards and cut them in two in the middle. For the front and back use twelve foot lumber and the waste pieces work in for roof boards if shingles are used.

ELEVATION

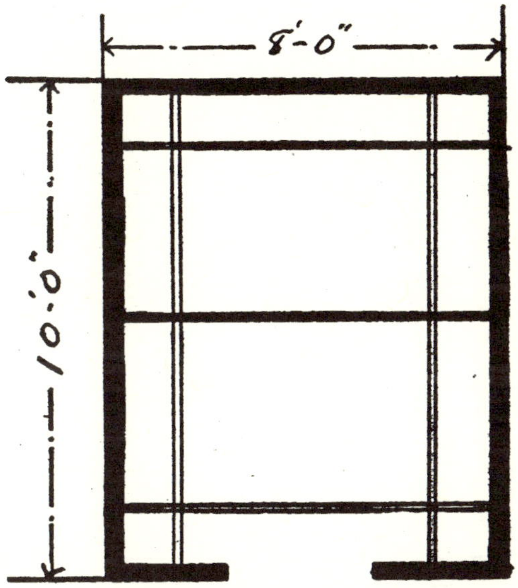

PLAN OF FLOOR

A smoke house like this is not tight enough to keep the meat in after being smoked. It is better to wrap it in paper, then roll it up in thin cotton and sew it up. You musn't leave a place for a fly to crawl in. You must then hang the packages with strings, perfectly free. They must not touch each other and they must not touch anything else. They need a cool place but not damp.

CEMENT BLOCK SMOKE HOUSE—A147

Every farm should have a smoke house, the better the house the more satisfactory will be the meat. The plans shown of (A147) is for a house constructed of cement blocks. It should be placed conveniently near the house on a raise of ground and a foundation started below the frost line. A trench should be dug, say 3½ feet deep partly filled with concrete made of one part of Portland cement, two and one-half parts sand and five parts of broken stone or gravel, ramming or puddling carefully. If plenty sand may be conveniently had, it would be a good plan to secure a block machine and have the blocks made on the ground. In making the concrete blocks, use a mixture of one part Portland cement, two and one-half parts sand and five parts of crushed stone or gravel. The use of crushed stone or coarse material for the back of the block saves a great deal of cement and at the same time gives a much better block than where sand and cement alone are used. Blocks made of sand and cement alone and merely dampened are not concrete blocks, but on the contrary are simply sand blocks. The very term of concrete suggests coarse material and plenty of water. Great care should be taken in mixing the different aggregates and they should be mixed thoroughly dry and after they have been thoroughly mixed add water. After the blocks have been made they should be set aside to be cured, and while curing, they should be sprayed thoroughly from seven to ten days. This spraying should commence about twelve hours after the block has been made. Blocks should never be used in building until they are from twenty to thirty days old.

Farm cured meats are a great luxury if the hogs are properly grown on pasture.

very carefully. The frame must have a couple of ridges all round and cement worked in tight between these ridges to make tight joints. The ventilator on top must be fitted with a fine screen. Two screens would be better. A coarse galvanized screen on top and a fine screen inside at the bottom.

The plates and rafters must be laid in fresh cement mortar on top of the wall. All spaces between rafters are filled in so as to prevent cracks or openings of any

ELEVATION

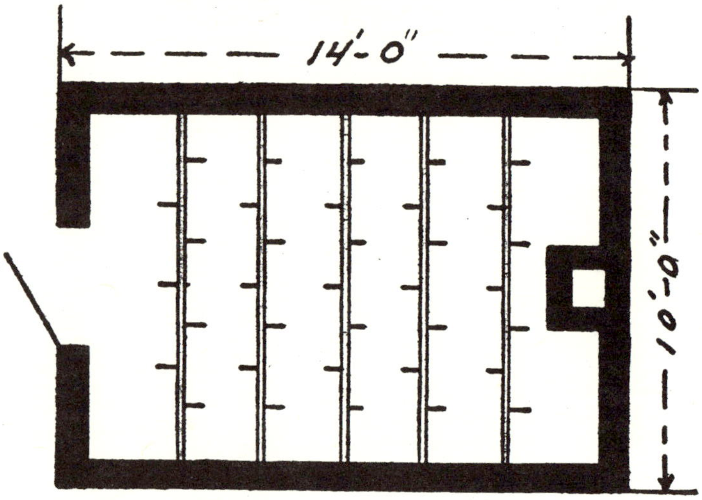

FLOOR PLAN

With a house like this and good pork to start with, a farmer can supply his table with good home-made bacon, hams and shoulders the year round.

The best smoke is made from green maple wood. Probably clean corn cobs come next. With a smoke house thoroughly well built to keep out flies and other insects the meat may be smoked in the spring and left in the smoke house all summer. By way of precaution a very little smoke may be started once or twice a month or some of the meat may be covered with paper and cloth. Very much depends on the house. If the house is too dry there will be too much evaporation and the meat will become dry; if the house is too damp it will be inclined to mould. If it is intended to keep the meat in the house after the smoking process is completed it will be necessary to fit the door

kind. Cross poles to support the meat are made of four by fours with half inch pegs inserted from the sides. The pegs are set at an angle of about thirty degrees. This will permit hanging the pieces of meat in the old fashioned way of cutting a slit in the skin in the bone end. If strings are preferred the same kind of peg may be used. Nails are not to be recommended for this purpose.

Farmers living within easy distance of a large city may work up a good trade in farmer cured meats by selling direct to consumers, thereby saving both the packer's and grocer's profits. Only thrifty young pigs, not too fat, and in prime condition are suitable for a trade of this kind, but farmers have the pick and they may just as well select the best. Every person living in the city enjoys a change in meats as well as in other things.

CHEAP GRAIN BUILDING—A128

A cheap building to hold grain and corn is shown in this design. It is a low building with studding only ten feet long, but that is about as high as a person cares to

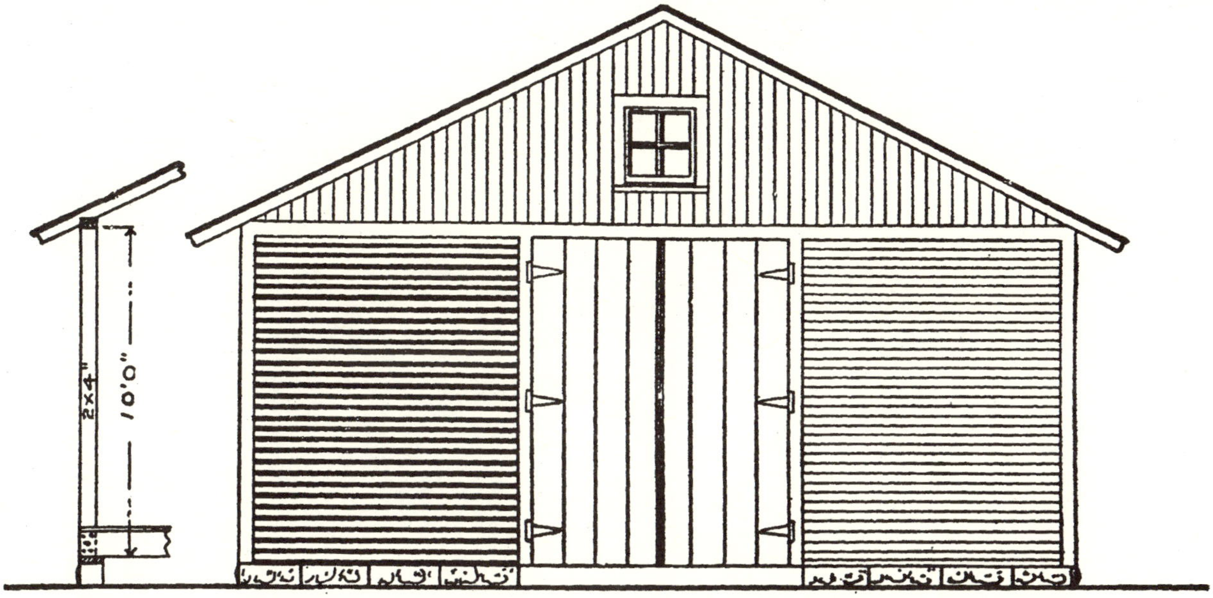

FRONT ELEVATION.

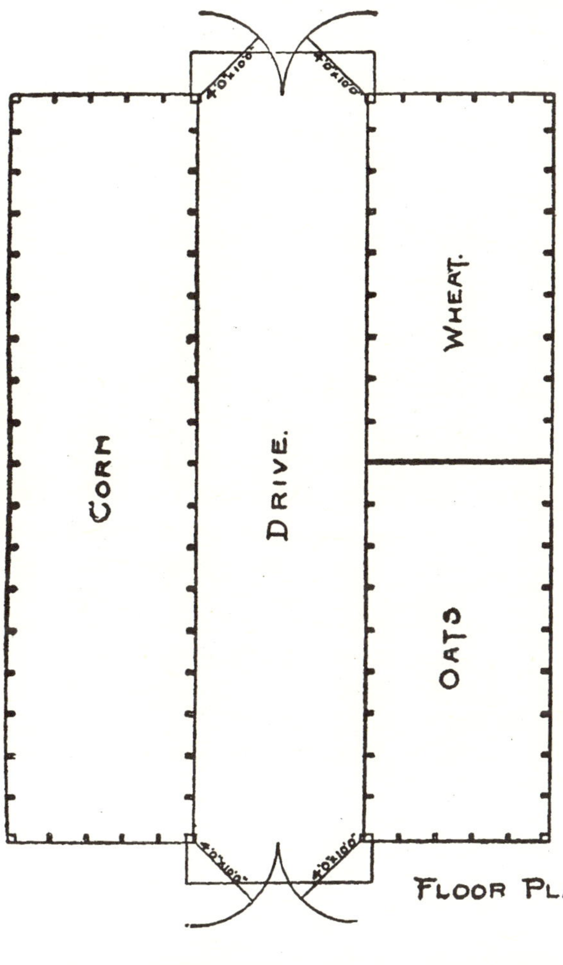

FLOOR PLAN.

pitch corn or threshed grain. Just ordinary one by four pine strips spaced to ¾ inch are nailed on the outside of two by four studding to make the corn crib, but the wheat and oat bins of course are made tight all around and a little extra work is put on the floor.

There is considerable side pressure in a wheat bin which must be guarded against by using a few extra braces, but heavy timbers are unnecessary in a bin the size of this one. This building may be floored overhead for storage, or the bins may be left open to the roof. By leaving the space open the building will be lighted sufficiently by the small window in each gable.

It is not intended to floor the driveway unless it is needed when using a fanning mill to clean grain, but the building would be all the better for having a good solid floor the full size. This plan provides for a building thirty by forty feet. Thirty feet is wide enough for convenience either in building or for use afterwards, but of course it may be any length.

GRANARY—A107

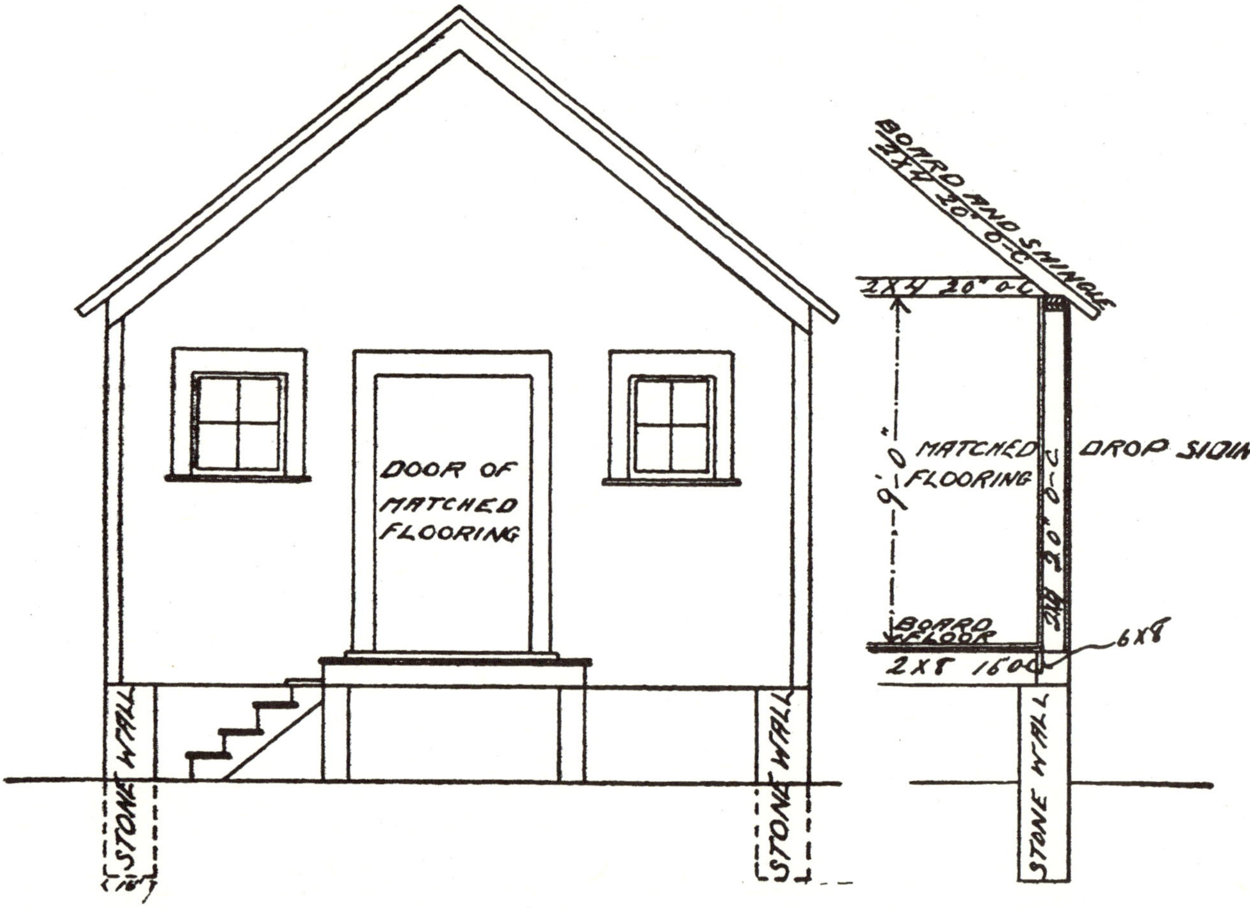

Farmers have more use for granaries than formerly. There are two reason for this, one is that more stock is kept on the farm and it is necessary to have grain the year round, another is that owing to a shortage of cars and speculation in grain, prices are not always satisfactory in the fall and it pays to hold grain to sell later.

Then, more attention is now being paid to seed. A grain house like this with a place for scales and a fanning mill is a very valuable addition to any farm. The different kinds of grain may be stored in the bins at threshing time and run through the fanning mill when taken to the warehouse for sale. By rigging the mill carefully a small proportion of the largest, heaviest grains may be retained for seed without adding anything to the cost. A good mill that will select say one bushel out of ten of the kind of grain that you want to sow and do it while blowing the chaff out of the grain you are selling without interfering with the grade is a valuable mill, but there are just such fanning mills made and their cost is little if any more than the common kind on the market.

In this scale room wires may be stretched for hanging the empty bags when not wanted. By sinking the scales in the floor each bag may be weighed as it is loaded. This is best done by having a two-wheeled bag truck and a counter weight on the scale beam so that the net weight may be written down each time without taking the time to calculate.

Great care should be taken in building

a granary to have it rat proof. The wall of course must go below the frost and it is a good plan to put a three inch tile all around the bottom on the outside which answers for drainage as well as to keep the house should be set up well from the ground for two reasons, it should be the height of the wagon for easy loading and unloading and it should be high and dry because grain should be kept from all un-

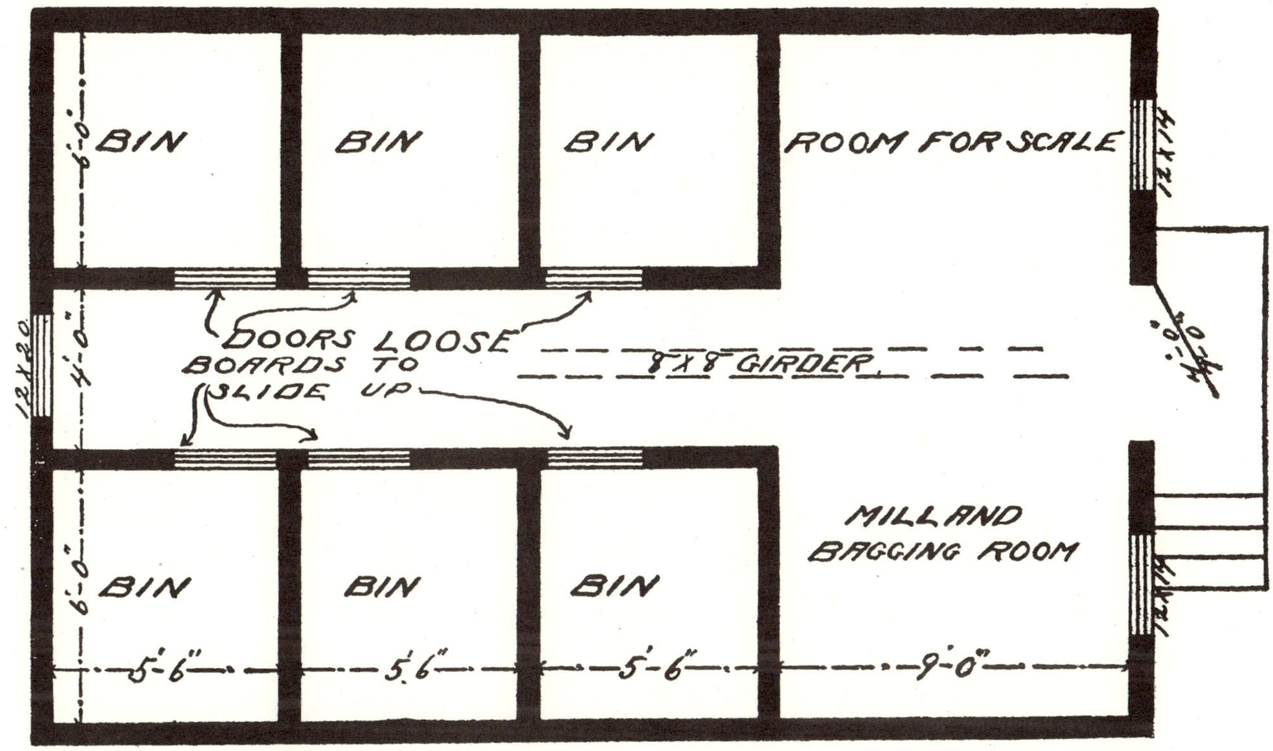

rats from burrowing under the wall. Some farmers object to a plaform in front of the door just on account of rats, but if the door is made heavy and made to fit tight with a bit of hoop iron at the bottom, rats will not get in that way if the door is kept shut. It is difficult to arrange a plan of getting in and out conveniently without a platform. The door is too high to step up and if you have a kind of stair to reach it you might just as well have a good loading platform as a cheap shaky affair. A grain necessary moisture. There is moisture enough in the air in damp weather anyhow without taking chances on moisture from the ground.

The doors to the bins are made of loose boards dropped into grooves so that one board may be put in or taken out as required. A little extra expense put into the quality of the flooring is money well laid out. The floor should be free from shake and fairly free from knots, at least there should be no black knots.

RAT PROOF GRANARY—A141

Cost of Blue Prints, $5.00

A dry floor and one that is ratproof is made by excavating for the foundation of the granary about six inches deep. Then pound in three or four inches of cinders and lay the sills and joists on the cinders.

After the building is up and enclosed make cement concrete by mixing one part cement, three parts sand and four parts gravel or broken stone. Fill in with this concrete to the tops of the joists, then while the concrete is soft put down the matched floor, nailing it right into the soft concrete. As soon as the floor is finished shut the building up tight and bank around the outside to keep the air away from the concrete so it will dry slowly.

The matched boarding is put on the outside of the studding and the siding nailed

END ELEVATION

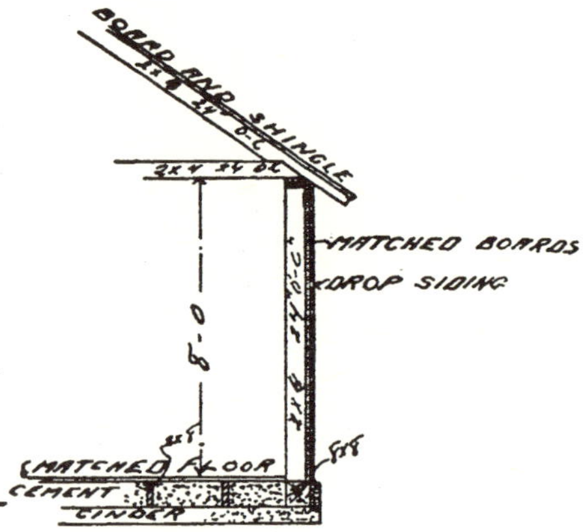

SECTION

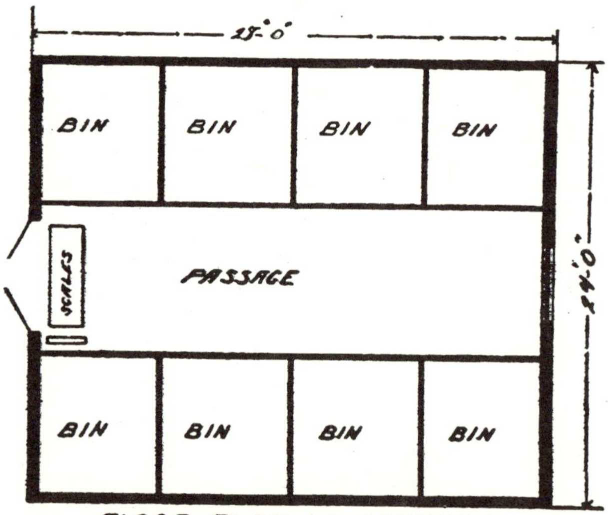

FLOOR PLAN OF GRANARY

over that. This is for the purpose of leaving the inside exposed so that a cat or dog could easily reach a rat if it should get inside. Hollow walls make harbors for rats but this construction leaves them no protection.

There is a window in the back end of the alley and another one over the door in front. The doors are made heavy and swing out. They close against heavy jambs so that rats and mice have very little encouragement to get in at the door. The scales are let in the floor flush. Provision must be made for this before the concrete is put in.

FARM HOUSE WATER SUPPLY UNDER PRESSURE—A226

It is just as easy and just as cheap to have a house water supply under pressure in the country as it is in the city because city rents and taxes are sufficient not only to pay for the cost but to make up for the stealings of dishonest municipal employees and the aggregate to the property holders foots up more than the interest on the cost of a sufficient private water supply system in the country.

We now have powerful windmills that will do the pumping and automatically attend to the work without very much supervision and with a very light annual expense for oil and repairs.

For extra large houses and where a great deal of stock is kept sometimes a gasoline engine is more satisfactory than a windmill but this is a question to be decided by local conditions because either one is all right when properly placed and connected.

The very first consideration is a good well. You must have plenty of water at all seasons and you must have pure water. It is more difficult to get pure water in the country than it is to get a plentiful supply

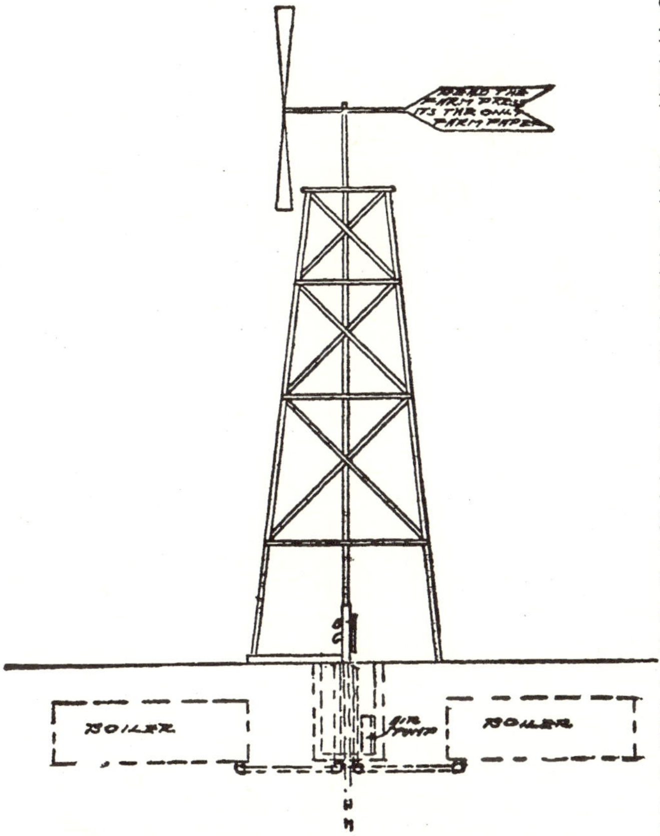

of water that is not especially good. The Illinois Sanitary Commission is responsible for the statement that most country wells contain disease germs, especially typhoid and dyphtheria.

There are two kinds of tanks in use; one is the ordinary elevated windmill tank that is in common use all over the country and the other is the underground pressure tank that has not yet attained universal popularity probably because it is new, but the underground tank has many advantages over the other. The water is cool in summer and it does not freeze in winter and the underground tank is cleaner because it is not open to dust and insects.

To buy a steel tank to put in the ground seems quite an expense to start with, but it often happens that an unsafe steam boiler may be bought very cheap for this purpose. A boiler maker will take out the old flues, patch the uncertain places and make the old shell perfectly water tight and satisfactory in every way.

The underground tank should be a little larger than necessary in order to have a safe supply of water at all times. Then you have the air space to figure on as this gives the pressure. You fill the tank about two-thirds or three-fourths full of water, then with an air pump you put on the pressure you want, which may be anywhere from ten to forty pounds per square inch. If your air pump is handy when needed the pressure may be quickly increased to one hundred pounds in case of fire which is another advantage of the underground tank system. By having a large tank and a large air space you get a steadier supply because the air pressure does not decrease so fast when you draw off water for use.

Every new country house should be piped with water and gas. It should be wired with electricity and the plumbing should be equal to any city residence. Farmers are entitled to all the improvements going. If they don't have them it is their own fault.

SCALE HOUSE—A187

This is a drawing of a good scale house covering an eight by fourteen foot platform, four ton scale. The building is fourteen by sixteen feet base with doors twelve feet high, allowing a large load of hay to be driven in upon the scales. The sides of the house are used for hanging and placing tools and other small articles not wanted in the barn.

Every up-to-date farm should have a good pair of scales big enough to weigh a load of hay or a drove of hogs or sheep. Enough money is lost on every farm by guessing at weights to pay for a good set of scales, and besides this there is a great satisfaction in knowing what things weigh.

In feeding cattle, hogs or sheep for market, weighing at regular periods is extremely valuable. It is impossible to know whether stock is doing as well as it

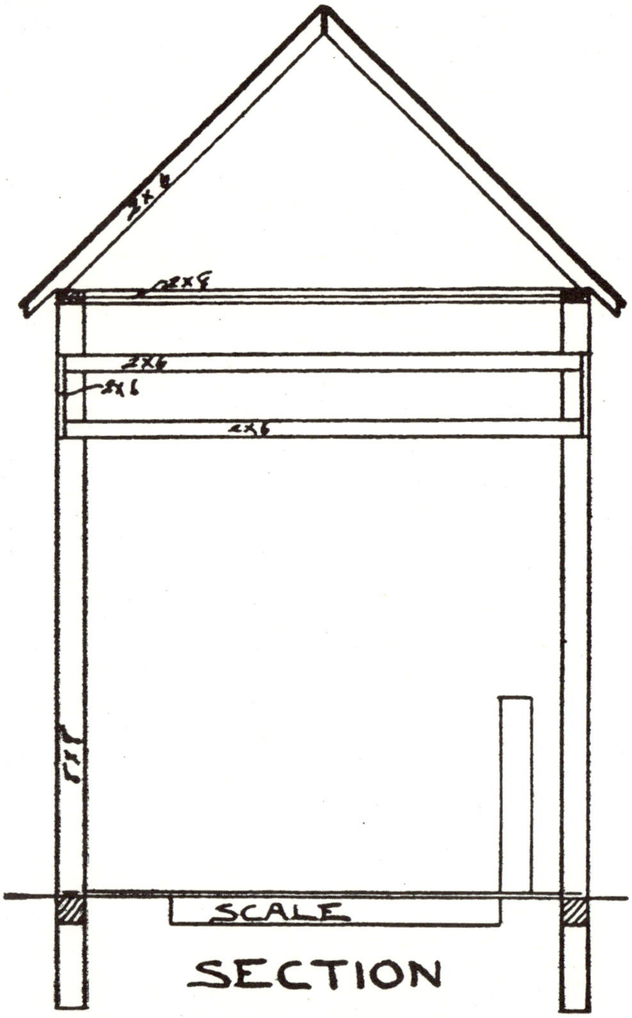

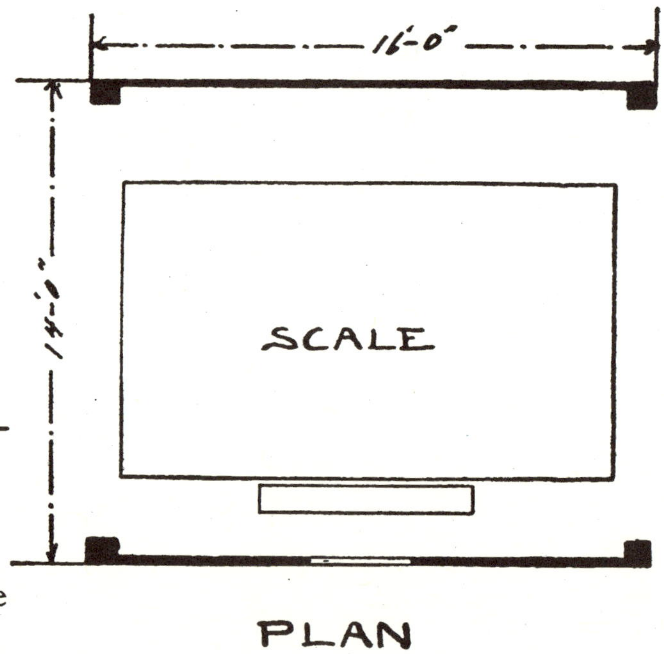

should do unless tab is kept on the increase in weight.

After this scale house is built make a good solid rack to surround the scales to pen up stock at weighing time. Make the rack high enough to hold horses, strong enough to hold a bull and tight enough for hogs. Wire fencing may be used to advantage, stretched on to wooden frames.

TOWER TANK HOUSE—A145

Where a water pressure is wanted it is often a good plan to put the water tank in the windmill tower. In plan (A145) the tank is shown in the dotted lines. It is placed ten feet above the ground and the tank itself is fourteen feet high by ten feet in diameter at the bottom.

In placing a tank like this it is necessary to carry a three inch pipe through the tank and pass the pump shaft through this pipe. The pipe is screwed into a flange at the bottom and the flange is bolted to the bottom of the tank to make it thoroughly water tight. The pipe must be steadied

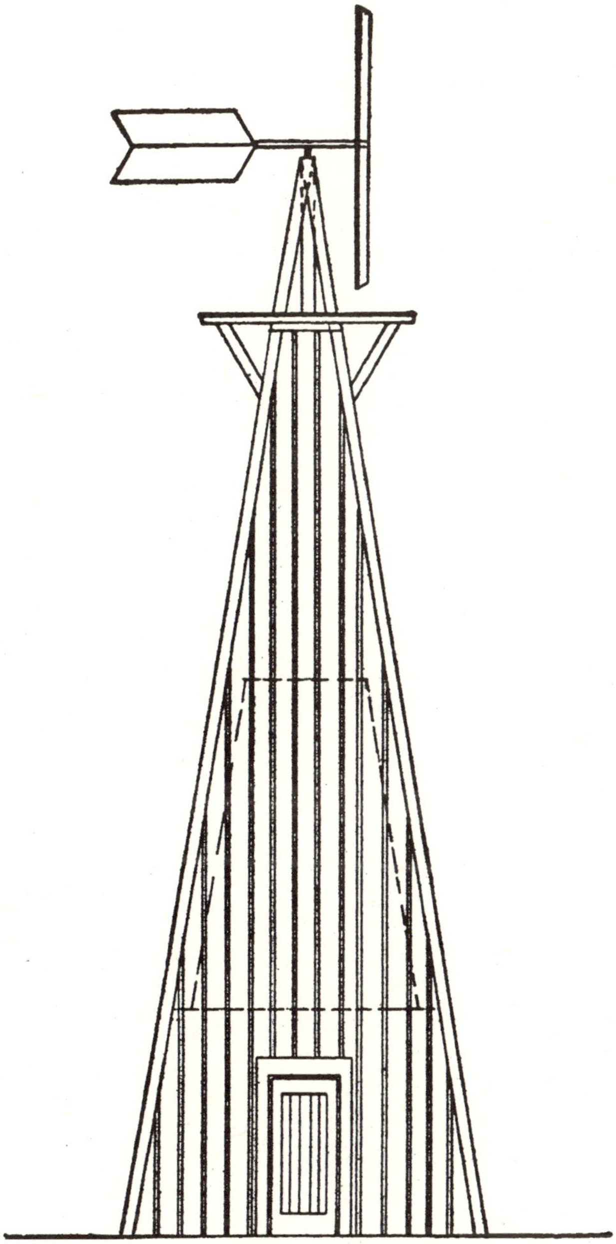

TANK HOUSE

at the top and the shaft must have a bearing, both above the tank and below it so it won't scrape on the pipe. The well and pump of course are directly under the tank in the center of the tower.

The outside boarding is made double and lined with paper to be warm in winter.

There is generally some drip from a tank placed like this for which reason the room below is seldom made use of for any purpose, but a few farmers have utilized this room for a bathroom. They make a cement bottom with a drain to carry off the surplus water and put in a shower bath connected with a pipe from the tank. A shower bath is the most convenient and probably the most healthful of any kind of a bath. At any rate it is easily kept clean.

There is no reason why the farmer or his men should be denied the privilege of getting a bath whey they want it. There are bathrooms in almost all city houses and there should be bathing conveniences on every farm. By placing a stove in this room under the tank it could be made comfortable in winter as well as summer, and a stove with a water heater attached to the tank would give a water pressure so that the shower could be made any temperature desired.

The height of this tower is forty feet to the windmill. Of course, the height of a windmill tower must depend upon its location. If the tower is built on high ground it is not necessary to go up so high unless the windmill is surrounded by high buildings or trees.

Index to Practical Barn Plans

Radford Building

1827-29-31-33 Prairie Ave., Chicago, Ill.

Publications by Algrove Publishing Limited

The following is a list of titles from our popular *"Classic Reprint Series"* as well as other publications by Algrove Publishing Limited.

ARCHITECTURE, BUILDING, AND DESIGN

Item #	Title
49L8096	☐ A GLOSSARY OF TERMS USED IN ENGLISH ARCHITECTURE
49L8137	☐ AUDELS CARPENTERS AND BUILDERS GUIDE - *VOLS. 1-4*
49L8016	☐ BARN PLANS & OUTBUILDINGS
49L8046	☐ BEAUTIFYING THE HOME GROUNDS
49L8112	☐ BUILDING WITH LOGS AND LOG CABIN CONSTRUCTION
49L8092	☐ DETAIL, COTTAGE AND CONSTRUCTIVE ARCHITECTURE
49L8015	☐ FENCES, GATES & BRIDGES
49L8706	☐ FROM LOG TO LOG HOUSE
49L8111	☐ LOW-COST WOOD HOMES
49L8030	☐ SHELTERS, SHACKS & SHANTIES
49L8139	☐ THE STAIR BUILDERS GUIDE
49L8050	☐ STRONG'S BOOK OF DESIGNS
49L8064	☐ THE ARCHITECTURE OF COUNTRY HOUSES
49L8023	☐ THE OPEN TIMBER ROOFS OF THE MIDDLE AGES

CLASSIC CATALOGS

Item #	Title
49L8004	☐ BOULTON & PAUL, LTD. 1898 CATALOGUE
49L8098	☐ CATALOG OF MISSION FURNITURE 1913 – *COME-PACKT FURNITURE*
49L8097	☐ MASSEY-HARRIS CIRCA 1914 CATALOG
49L8089	☐ OVERSHOT WATER WHEELS FOR SMALL STREAMS
49L8079	☐ WILLIAM BULLOCK & CO. – *HARDWARE CATALOG, CIRCA 1850*

GARDENING

Item #	Title
49L8082	☐ CANADIAN WILD FLOWERS (C. P. TRAILL)
49L8113	☐ COLLECTING SEEDS OF WILD PLANTS AND SHIPPING LIVE PLANT MATERIAL
49L8029	☐ FARM WEEDS OF CANADA
49L8056	☐ FLORA'S LEXICON
49L8705	☐ REFLECTIONS ON THE FUNGALOIDS
49L8057	☐ THE WILDFLOWERS OF CANADA

HUMOR AND PUZZLES

Item #	Title
49L8074	☐ ARE YOU A GENIUS? WHAT IS YOUR I.Q?
49L8106	☐ CLASSIC COWBOY CARTOONS, VOL. 1 (J.R. WILLIAMS)
49L8109	☐ CLASSIC COWBOY CARTOONS, VOL. 2 (J.R. WILLIAMS)
49L8118	☐ CLASSIC COWBOY CARTOONS, VOL. 3 (J.R. WILLIAMS)
49L8119	☐ CLASSIC COWBOY CARTOONS, VOL. 4 (J.R. WILLIAMS)
49L8072	☐ CLASSIC PUZZLES AND HOW TO SOLVE THEM
49L8103	☐ GRANDMOTHER'S PUZZLE BOOK 1
49L8142	☐ GRANDMOTHER S PUZZLE BOOK 2
49L8127	☐ JOIN THE DOTS PUZZLE BOOKS
49L8081	☐ MR. PUNCH WITH ROD AND GUN – *THE HUMOUR OF FISHING AND SHOOTING*
49L8073	☐ NAME IT! THE PICTORIAL QUIZ BOOK
49L8126	☐ OUR BOARDING HOUSE WITH MAJOR HOOPLE – *1927*
49L8125	☐ OUT OUR WAY–*SAMPLER 20s, 30s & 40s* (J.R. WILLIAMS)
49L8044	☐ SAM LOYD'S PICTURE PUZZLES
49L8084	☐ THE ART OF ARTHUR WATTS
49L8071	☐ THE BULL OF THE WOODS, VOL. 1 (J.R. WILLIAMS)
49L8080	☐ THE BULL OF THE WOODS, VOL. 2 (J.R. WILLIAMS)
49L8104	☐ THE BULL OF THE WOODS, VOL. 3 (J.R. WILLIAMS)
49L8114	☐ THE BULL OF THE WOODS, VOL. 4 (J.R. WILLIAMS)
49L8115	☐ THE BULL OF THE WOODS, VOL. 5 (J.R. WILLIAMS)
49L8116	☐ THE BULL OF THE WOODS, VOL. 6 (J.R. WILLIAMS)
49L8610	☐ THE CLASSIC ART OF HAND SHADOWS
49L8612	☐ THE FLIGHT OF THE BUMBLE BEE
49L8128	☐ THE NIGHT BEFORE CHRISTMAS WITH PUZZLE PICTURES
49L8107	☐ U.S. CAVALRY CARTOONS (J.R. WILLIAMS)

NAVAL AND MARINE

Item #	Title
49L8090	☐ BOAT-BUILDING AND BOATING
49L8707	☐ BUILDING THE NORWEGIAN SAILING PRAM *(MANUAL AND PLANS)*
49L8708	☐ BUILDING THE SEA URCHIN *(MANUAL AND PLANS)*
49L8138	☐ HOW SAILS ARE MADE AND HANDLED
49L8078	☐ MANUAL OF SEAMANSHIP FOR BOYS AND SEAMEN OF THE ROYAL NAVY, 1904
49L8129	☐ OLD SHIP FIGURE-HEADS & STERNS
49L8095	☐ SAILING SHIPS AT A GLANCE
49L8134	☐ SAILING VESSEL SILHOUETTES
49l8144	☐ THE KEDGE-ANCHOR
49L8099	☐ THE SAILOR'S WORD-BOOK
49L8605	☐ THE SAILOR S POCKET BOOK OF KNOTS
49L8058	☐ THE YANKEE WHALER
49L8025	☐ THE YOUNG SEA OFFICER'S SHEET ANCHOR
49L8061	☐ TRADITIONS OF THE NAVY

REFERENCE

Item #	Title
49L8024	☐ 1800 MECHANICAL MOVEMENTS AND DEVICES
49L8093	☐ 507 MECHANICAL MOVEMENTS
49L8055	☐ 970 MECHANICAL APPLIANCES AND NOVELTIES OF CONSTRUCTION
49L8602	☐ ALL THE KNOTS YOU NEED
49L8083	☐ AMERICAN MECHANICAL DICTIONARY – KNIGHT VOL. I, VOL. II, VOL. III
49L8077	☐ CAMP COOKERY
49L8711	☐ DESIGNING SUNDIALS – *THE GRAPHIC METHOD*
49L8145	☐ HAWKINS MECHANICAL DICTIONARY
49L8001	☐ LEE'S PRICELESS RECIPES
49L8135	☐ MUSSON S IMPROVED LUMBER AND LOG POCKET BOOK
49L8018	☐ THE BOY'S BOOK OF MECHANICAL MODELS
49L8610	☐ THE CLASSIC ART OF HAND SHADOWS
49L8019	☐ WINDMILLS AND WIND MOTORS

TRADES

Item #	Title
49L8014	☐ BOOK OF TRADES
49L8086	☐ FARM BLACKSMITHING
49L8031	☐ FARM MECHANICS
49L8141	☐ FARM WORKSHOP GUIDE
49L8087	☐ FORGING
49L8027	☐ HANDY FARM DEVICES AND HOW TO MAKE THEM
49L8002	☐ HOW TO PAINT SIGNS & SHO' CARDS
49L8054	☐ HOW TO USE THE STEEL SQUARE
49L8094	☐ THE YOUNG MILL-WRIGHT AND MILLER'S GUIDE

WOODWORKING AND CRAFTS

Item #	Title
49L8130	☐ 50 POPULAR WOODWORKING PROJECTS
49L8012	☐ BOY CRAFT
49L8110	☐ CHAIN SAW AND CROSSCUT SAW TRAINING COURSE
49L8048	☐ CLAY MODELLING AND PLASTER CASTING
49L8005	☐ COLONIAL FURNITURE
49L8065	☐ COPING SAW WORK
49L8032	☐ DECORATIVE CARVING, PYROGRAPHY AND FLEMISH CARVING
49L8091	☐ FURNITURE DESIGNING AND DRAUGHTING
49L8049	☐ HANDBOOK OF TURNING
49L8020	☐ MISSION FURNITURE, HOW TO MAKE IT
49L8710	☐ QUEEN ANNE FURNITURE - *HISTORY, DESIGN & CONSTRUCTION*
49L8003	☐ RUSTIC CARPENTRY
49L8085	☐ SKELETON LEAVES AND PHANTOM FLOWERS
49L8068	☐ SPECIALIZED JOINERY
49L8052	☐ STANLEY COMBINATION PLANES –*THE 45, THE 50 & THE 55*
49L8034	☐ THE ART OF WHITTLING
49L8131	☐ TIN-CAN PROJECTS AND ART-METAL WORK
49L8042	☐ TURNING FOR AMATEURS
49L8067	☐ WOOD HANDBOOK – *WOOD AS AN ENGINEERING MATERIAL*
49L8060	☐ WOODEN PLANES AND HOW TO MAKE THEM
49L8013	☐ YOU CAN MAKE IT
49L8035	☐ YOU CAN MAKE IT FOR CAMP & COTTAGE
49L8036	☐ YOU CAN MAKE IT FOR PROFIT

Algrove Publishing Limited, 36 Mill Street, P.O. Box 1238, Almonte, Ontario, Canada K0A 1A0
Telephone: (613) 256-0350 Fax: (613) 256-0360 Email: sales@algrove.com Web: www.algrove.com